DOLPHINS
AND THE
TUNA INDUSTRY

Committee on Reducing Porpoise Mortality from Tuna Fishing
Board on Biology
Board on Environmental Studies and Toxicology
Commission on Life Sciences
National Research Council

NATIONAL ACADEMY PRESS
Washington, D.C. 1992

NATIONAL ACADEMY PRESS • 2101 Constitution Avenue, NW • Washington, DC 20418

NOTICE: The project that is the subject of this report was approved by the Governing Board of the National Research Council, whose members are drawn from the councils of the National Academy of Sciences, the National Academy of Engineering, and the Institute of Medicine. The members of the committee responsible for the report were chosen for their special competencies and with regard for appropriate balance.

This report has been reviewed by a group other than the authors according to procedures approved by a Report Review Committee consisting of members of the National Academy of Sciences, the National Academy of Engineering, and the Institute of Medicine.

Support for this project was provided by the U.S. Department of Commerce, NOAA Contract No. 50-DGNF-9-00157.

Cover photo: Courtesy, NOAA Photo Library

Library of Congress Cataloging-in-Publication Data

National Research Council (U.S.). Committee on Reducing Porpoise Mortality from Tuna Fishing.
    Dolphins and the tuna industry/Committee on Reducing Porpoise Mortality from Tuna Fishing, Board on Biology, Board on Environmental Studies and Toxicology, Commission on Life Sciences, National Research Council.
        p.   cm.
    Includes bibliographical references and index.
    ISBN 9-309-04735-8
    1. Tuna fisheries—Environmental aspects. 2. Dolphins—Mortality. 3. Tuna industry—Environmental aspects. I. Title.
    SH351.T8N37   1992
    333.95′9—dc20   92-10603
                                                  CIP

Committee on Reducing Porpoise Mortality from Tuna Fishing

ROBERT C. FRANCIS, *Chairman*, University of Washington, Seattle
FRANK T. AWBREY, San Diego State University and Hubbs Sea World
Research Institute, San Diego
CLIFFORD A. GOUDEY, Massachusetts Institute of Technology, Cambridge
MARTIN A. HALL, Inter-American Tropical Tuna Commission,
La Jolla, CA
DENNIS M. KING, University of Maryland, Solomons
HAROLD MEDINA, Jamul, CA
KENNETH S. NORRIS, University of California, Santa Cruz
MICHAEL K. ORBACH, East Carolina University, Greenville, NC
ROGER PAYNE, Long Term Research Institute, Lincoln, MA
ELLEN PIKITCH, University of Washington, Seattle

Staff

DAVID POLICANSKY, Project Director, 1990–1992
DAVE JOHNSTON, Project Director, 1989–1990
RUTH CROSSGROVE, Editor
SHIRLEY JONES, Project Assistant
BERNIDEAN WILLIAMS, Information Specialist

Preface

The Committee on Porpoise Mortality from Tuna Fishing was formed on October 10, 1989, under the auspices of the Board on Biology (BB) and the Board on Environmental Studies and Toxicology (BEST) of the National Research Council's Commission on Life Sciences. The committee was formally charged as follows:

The task of this committee is to perform a study mandated by the Marine Mammal Protection Act Amendments of 1988, reviewing scientific and technical information relevant to promising new techniques for finding and catching yellowfin tuna without killing porpoises in the Eastern Tropical Pacific Ocean. The committee will review information on the biology and ecology of the yellowfin tuna and porpoises with which they commonly associate, as well as the nature of the "tuna-porpoise bond." The committee will also identify currently available and promising new techniques for reducing the incidental catch of porpoises in tuna fishing. These techniques include locating tuna without porpoises, breaking the tuna-porpoise bond, and modifying gear or methods to reduce incidental drowning of porpoises in nets. The resulting report will be used by the Secretary of Commerce as a basis for a proposed plan for research, development, and implementation of alternative fishing techniques.

The committee met on 4–5 December 1989, 1–3 February 1990, 21–22 April 1990, and 11–12 June 1990. In addition, a writing group of R. Francis and M. Orbach met in Morehead City, N.C., in July 1990.

The committee was fortunate to have in its membership experts on both tuna and dolphin biology and ecology, tuna fishing and gear technology (in particular, purse-seine and dolphin-conservation technology), fishery biology, management, sociology, and economics. All of these kinds of expertise were required to address the complex issues of dolphin mortality from tuna fishing. Indeed, the problem goes far beyond issues of science and technology.

In the early 1970s, the "tuna-dolphin problem" was one of the most difficult issues before the newly formed Marine Mammal Commission. Now, two decades later, it still remains a matter of intense concern to politicians, the media, and the public at large. This concern persists not because of its scientific importance, but because its ramifications extend into many aspects of contemporary society. It affects not only those directly involved in the harvesting and marketing of tuna and tuna products, but also the competitive relationship between U.S. and non-U.S. fishermen as well as the international relations between the United States and other countries that harvest tuna and market tuna products. Perhaps even more important, the tuna-dolphin problem raises a complex series of sociological, and inevitably political, issues that include attitudes of and about fishermen, food processors and canners, retailers, animal-rights advocates, disparate groups of environmental activists, the popular press, and other news media.

The committee, however, was charged to identify scientific and technological innovations that show promise in reducing dolphin mortality from tuna fishing. Although the committee decided that the problem must be put into the broad context of policy, history, and economics, it strove to provide an unbiased, neutral evaluation of the scientific and technical issues it faced. Individual members hold personal views on the values of preventing all dolphin mortality from tuna fishing, of preserving the fishermen's way of life, of obtaining valuable protein at the lowest possible cost, and of other related issues, but the committee made every effort to exclude those personal views from its analysis and from this report. The committee's effort was focused on providing, to the best of its ability, the best scientifc and technical evaluation to inform policymakers and the public, even if some of its scientific conclusions are unpalatable to various people.

The committee was assisted greatly by the National Research Council staff, in particular our project directors Dave Johnston and David Policansky, our project assistants Linda Kegley and Shirley Jones, and editor Ruth Crossgrove. Their work was essential in producing a coherent report from the drafts provided by the committee members. We also acknowledge the guidance of BEST liaison Joanna Burger and BB liaison David Wake. We are grateful to nine anonymous reviewers and to George Bartholomew for contributing many helpful comments and suggestions.

We also thank persons who provided information, prepared presentations for the committee, or made original data available to the committee for analysis and interpretation. They include David Bratten, David Cormany, Doug DeMaster, Rick Deriso, Robert Hofman, Ken Hollingshead, Michael Laurs, Richard McNeely, Charles Oliver, William Perrin, Karen Pryor, Joyce Sisson, Pat Tomlinson, and John Twiss.

I wish to thank two graduate students at Fisheries Research Institute, James Ianelli and Alejandro Anganuzzi, who assisted with my analysis and interpretation of data. Finally, I also wish to thank four outstanding individuals who helped to make my task a doable one. Abby Simpson took care of all matters of logistics and editing at the FRI end. Her keen knowledge of the English language greatly enhanced the quality of the sections of the report drafted by me. Committee members Martin Hall and Harold Medina have consistently acted beyond the call of duty. Martin has patiently provided the committee with more information and a greater critical eye than I would have thought possible. Harold, as the only committee member from the U.S. tuna industry, has consistently provided his unique insight, sometimes in very contentious and uncomfortable situations, in the most distinguished and gentlemanly fashion imaginable. And last but not least, in addition to providing technical and editorial support, Program Director David Policansky has been a spiritual companion throughout the course of this long and difficult journey.

Robert C. Francis
Chairman

The National Academy of Sciences is a private, nonprofit, self-perpetuating society of distinguished scholars engaged in scientific and engineering research, dedicated to the furtherance of science and technology and to their use for the general welfare. Upon the authority of the charter granted to it by the Congress in 1863, the Academy has a mandate that requires it to advise the federal government on scientific and technical matters. Dr. Frank Press is president of the National Academy of Sciences.

The National Academy of Engineering was established in 1964, under the charter of the National Academy of Sciences, as a parallel organization of outstanding engineers. It is autonomous in its administration and in the selection of its members, sharing with the National Academy of Sciences the responsibility for advising the federal government. The National Academy of Engineering also sponsors engineering programs aimed at meeting national needs, encourages education and research, and recognizes the superior achievements of engineers. Dr. Robert M. White is president of the National Academy of Engineering.

The Institute of Medicine was established in 1970 by the National Academy of Sciences to secure the services of eminent members of appropriate professions in the examination of policy matters pertaining to the health of the public. The Institute acts under the responsibility given to the National Academy of Sciences by its congressional charter to be an adviser to the federal government and, upon its own initiative, to identify issues of medical care, research, and education. Dr. Kenneth I. Shine is president of the Institute of Medicine.

The National Research Council was organized by the National Academy of Sciences in 1916 to associate the broad community of science and technology with the Academy's purposes of furthering knowledge and advising the federal government. Functioning in accordance with general policies determined by the Academy, the Council has become the principal operating agency of both the National Academy of Sciences and the National Academy of Engineering in providing services to the government, the public, and the scientific and engineering communities. The Council is administered jointly by both Academies and the Institute of Medicine. Dr. Frank Press and Dr. Robert M. White are chairman and vice chairman, respectively, of the National Research Council.

Contents

DOLPHINS
AND THE
TUNA INDUSTRY

Executive Summary

INTRODUCTION

Approximately one quarter of the world's tuna catch is taken in the eastern tropical Pacific Ocean (ETP). In that area, the most economically important tuna species, the yellowfin *(Thunnus albacares)*, is often found in association with various species of dolphins. Increasingly since the late 1940s, tuna fishermen have taken advantage of this association and have caught tuna by setting their nets around the highly visible herds of dolphins, which, being mammals, must surface often to breathe. Despite improvements in techniques and in gear that have substantially reduced the number of dolphins killed in the ETP tuna fishery, thousands of dolphins are still killed each year.

When the Congress reauthorized the Marine Mammal Protection Act of 1972 (MMPA) on November 23, 1988, it stipulated several amendments to the act. One of these amendments (Section 110(a)) focused on identifying appropriate research into promising new methods of locating and catching yellowfin tuna without the incidental capture of dolphins. It further directed the Secretary of Commerce to arrange for "an independent review of information pertaining to such potential alternative methods to be conducted by the National Academy of Sciences with individuals having scientific, technical or other expertise that may be relevant to the identification of promising alternative fishing techniques." The report from the National Academy of Sciences would then be submitted by the Secretary, together with a proposed plan of research, development, and implementation of alternative fishing techniques, to the Committee on Commerce, Science, and Transportation of the Senate and the Committee on Merchant Marine and Fisheries of the House of Representatives.

This report, by the National Research Council's Committee on Reducing Porpoise Mortality from Tuna Fishing, provides background information on catching yellowfin tuna and the incidental capture of dolphins in the ETP, recommendations from workshops focusing on improving current fishing techniques, and recommendations for research on alternative techniques.

THE PROBLEM

Catching tuna with purse seines is the most efficient method currently available. There are three major modes of purse-seine fishing. The first is school fishing, in which schools of tuna near the surface are found visually and the purse seine is set around the school. The second, called log fishing, depends on the attractive power of floating objects for tuna. Purse seines are set around natural logs or fish-aggregating devices to catch the fish that are often associated with them. The third method—the main focus of this report—is "dolphin fishing." Yellowfin tuna and certain species of dolphins are often associated, especially in the ETP. Dolphins are easy to see from a boat because of their frequent surfacing. The purse seine is set around the dolphin herd, and because the tuna are closely associated with them, catching dolphins usually means catching tuna as well. Although a variety of techniques and equipment have been developed to release the dolphins safely, thousands are still killed each year by dolphin fishing.

Although all three methods of purse seining catch tuna, log and school fishing catch mostly small, sexually immature tuna. Dolphin fishing usually catches large fish that are often sexually mature and produces larger average catches than the other methods. Thus, redirecting the fishing away from tuna associated with dolphins would be less efficient and would have a negative effect on the yield of the fishery and perhaps on the conservation of tuna populations. Considering only the point of view of economics and harvesting tuna, large tuna should be sought; fishermen should fish on dolphins and be discouraged from fishing on logs or schools. However, dolphin fishing kills dolphins; to minimize the killing of dolphins, all fishing should be directed away from dolphins. This dichotomy is the basis of the tuna-dolphin problem.

Two operational factors are related to the total number of dolphins killed in purse seining for tuna. The first is the number of times purse seines are set around dolphins, and the second is the number of dolphins killed in each such set. The first factor is affected by market prices, the availability of tuna of different sizes, restrictions such as the policy of some processors not to buy tuna caught on trips that have involved the intentional encirclement or death of dolphins (so-called "dolphin-safe" tuna), and the availability of alternative methods of catching tuna. The second factor depends on conditions such as the fishermen's motivation, skill, and experience; the condition of the vessel's equipment; weather conditions; and technological developments. Therefore, two general approaches are available to achieve the goal of reducing dolphin mortality—economic and technical—and each approach has several options.

The tuna-dolphin problem has several components. First, it is an international problem. Part of the difficulty is due to differing conservation ethics—U.S. laws and policies have the goal of preventing all dolphin mortality from tuna fishing, whereas the laws and policies of other nations are more often directed toward conserving dolphin populations, but not necessarily preventing all mortality. The international scope of the problem was illustrated by a 1991 ruling of a panel of judges that a U.S. embargo imposed in 1990 on Mexican tuna—applied in accordance with the MMPA—violates the provisions of the General Agreement on Tariffs and Trade.

Economic factors also are important. The U.S. tuna fleet in the ETP has become very small, probably because of a variety of economic factors, probably including the recent decision by major canneries to buy only "dolphin-safe" tuna. As a result, regulations on the U.S. fleet and adoption by the U.S. fleet of new gear and techniques might not be effective in further reducing dolphin mortality unless they are supported to some degree by other countries and thus affect the activities of non-U.S. boats as well as U.S. boats. The effects of decisions with respect to buying only certain kinds of tuna will depend on the U.S. share of the market and other factors. Indeed, if the recent rate of its decline continues, the U.S. fleet will soon cease to play a significant role in fishing for dolphin-associated tuna in the ETP.

The committee's analysis made it clear that no methods of catching tuna without killing dolphins—currently available or capable of rapid development—are as efficient as current methods of catching large yellowfin tuna in the ETP. Therefore, although it was not specifically in its charge, the committee focused on modifications of current methods, as well as identifying areas for research that might lead to new methods several years down the road. The committee made no attempt to evaluate the relative costs and benefits of such a research program.

THE CURRENT FISHERY

More than 70 nations participate in the world tuna fishery, but only 10 of those nations account for almost 85% of the catch. In 1989, Japan accounted for about 29% and the United States for 12%. About 36% of the total catch is consumed by Japan and about 31% by the United States. Over the years, the ETP has been one of the most productive tuna fishing areas in the world.

The ETP fishery was formerly dominated by the U.S. fleet. Since 1960, the fleets of nations other than the United States have increased and the U.S. fleet has decreased, especially in recent years. The proportion of the catch taken by

U.S. vessels decreased from 90% in 1960 to 32% in 1988 to 11% in 1991, and that taken by Latin American countries bordering the Pacific Ocean increased from 10% to 47% in 1988 and 57% by the end of 1991.

Until about 1975, the U.S. market consumed about 80% of the catch of surface-caught tuna from the ETP. With the expansion of the non-U.S. fleets during the 1980s, the share of the catch absorbed by the U.S. market declined, falling to 45% in 1987. Most of the difference went to Latin America, Europe, and the Far East.

Of particular interest to the committee is the surface fishery (bait-pole and purse seine) for yellowfin tuna in the ETP. Surface-caught yellowfin and skipjack tuna provide the raw material for the canned light-meat tuna product marketed primarily in the United States, Europe, and the Far East. Over the years, as new fishing areas have become viable, the ETP has contributed a smaller fraction of the world catch of yellowfin and skipjack.

DOLPHIN MORTALITY ASSOCIATED WITH THE TUNA FISHERY

Although no accurate data on dolphin mortality are available for the early years (1950–1972), the increased offshore operation of increasingly sophisticated fishing vessels setting their nets on herds of dolphins clearly led to very high mortality, especially in the ETP. Even though published estimates of dolphin mortality from 1960 to 1972 are based on a small and probably biased data set, and although such estimates have varied among authors and from year to year, it seems likely that more than 100,000 dolphins were killed annually by the U.S. fleet. After 1972, the annual number killed by the U.S. fleet declined to approximately 20,000 in 1979 and an estimated 19,712 in 1988 and 12,643 in 1989. This decline was due to a decrease in the rate at which tuna boats killed dolphins and a decrease in the number of boats in the U.S. fleet. Mortality is much larger among the tuna fleets registered in increasing numbers in other countries. By 1986, only 34 boats of the 103-boat fleet were registered in the United States and subject to National Marine Fisheries Service (NMFS) regulations; by 1991, only 11 U.S. boats fished the ETP. The Inter-American Tropical Tuna Commission (IATTC) estimates that the total kill for 1989 was 97,000–102,000 dolphins, of which the kill of 12,643 attributed to the U.S. fleet was less than 15%. In 1990, total mortality declined to 52,000–56,000 dolphins, of which less than 10% (5,083 dolphins) was attributed to the U.S. fleet. In 1991, average mortality per set has continued to decline and is now close to half of the 1990 values. Fishing effort on dolphins has also declined, and the total mortality for 1991 may be around 25,000 dolphins.

Three species of dolphins are most commonly affected: the spotted dolphin *(Stenella attenuata)*, the spinner dolphin *(Stenella longirostris)*, and the common dolphin *(Delphinus delphis)*. NMFS, which has responsibility for protecting marine mammals under the MMPA, set an annual quota in 1980 of 20,500 dolphins that could be killed by U.S. tuna boats, that quota including subquotas for various species and stocks. Additional regulations apply to fishing techniques.

Improvements in fishing gear and techniques discussed in Chapter 7, the mandatory observer program, and the kill quota were responsible for the dramatic decline in the number of dolphins killed by boats in the U.S. fleet. Currently, only about 0.5% of dolphins encircled by purse seines are injured or killed, and almost all of these are killed in a small number of problem sets. Most sets on dolphins are now zero-kill sets.

Most major modifications of gear and techniques were implemented more than a decade ago. More recent declines in mortality can be traced to improvements in equipment and the performance of the fishermen. Industrial groups have set up national technical advisory offices that, in collaboration with the IATTC, include organizing training courses for vessel captains and crews, inspecting the condition and performance of dolphin-saving gear, and diagnosing individual vessel or skipper problems. The recent declines in mortality, from 130,000 dolphins in 1986 to perhaps 25,000 in 1991, were achieved without any major technological advances or additions to the fishing gear.

Although mortality of dolphins has been drastically reduced, the annual kill is still substantial. In addition to the humanitarian and ecological concern for dolphins, the killing of dolphins is detrimental to fishermen. Dolphins are viewed as a valuable resource to fishermen because they attract and hold large yellowfin tuna near the surface where they can be caught. Thus, reducing dolphin mortality as much as possible is in the interest of the fishermen.

U.S. TUNA AND MARINE MAMMAL POLICY AND ECONOMICS

The U.S. government has been involved in some form of tuna-policy negotiations since the 1940s and in formal regulatory activity concerning tuna fisheries since the 1960s. The relevant laws and activities include the founding of IATTC in 1949 and the Tuna Conventions Act of 1950, the Fishermen's Protective Act (1954), the International Commission on the Conservation of Atlantic Tunas and the Atlantic Tuna Conventions Act (1975), the U.S.-Canada Pacific Albacore Treaty (1982), and the South Pacific Tuna Act (1988). IATTC and the Tuna Conventions Act of 1950 and the Fishery Conservation and Management Act of 1976 have had the greatest impact on the tuna industry.

In addition to the general U.S. tuna-policy framework, the U.S. policies on marine mammal conservation and protection have significantly affected the U.S. high-seas tuna fleet. The most significant of these policies result from the MMPA (Chapter 2). The act prohibits the taking of any marine mammal. It

(Section 1371) was amended in 1981 to state that " ... [the] goal [of zero mortality] shall be satisfied in the case of the incidental taking of marine mammals in the course of purse-seine fishing for yellowfin tuna by a ˙continuation of the application of the best marine mammal safety techniques and equipment that are economically and technologically practicable."

The tuna fishery is part of a highly capitalized and international food-processing industry. This situation, combined with the migratory nature of tuna, results in complex policy and management conditions that are critical to any effort to preserve and conserve dolphin populations.

The U.S. tuna market is 31% of the global tuna market, and an estimated 70% of the U.S. tuna supply is imported in either raw/frozen or canned form. U.S. tuna harvesters and processors have been moving away from the United States, leaving it more dependent on tuna imports. This trend will probably accelerate if any new costs are imposed on U.S. tuna harvesters or processors that are not incurred equally by foreign tuna companies. To the extent that attempts by the United States to reduce dolphin mortality increase costs or reduce productivity for U.S. vessels, these vessels will lose competitive ability in the ETP tuna fishery and are likely to be sold to foreign investors to remain competitive. Whether the U.S. government can (1) prevent the loss of U.S. vessels, (2) require U.S. tuna fishermen to operate at a competitive disadvantage, (3) subsidize fishermen to remain under U.S. jurisdiction, or (4) remove the advantage of foreign vessels by restricting access to the U.S. market remains unclear. Developing engineering-based solutions to the tuna-dolphin problem should be viewed as only the first step in reducing dolphin mortality. Unless the solution is cost-effective, U.S. and foreign tuna harvesters are unlikely to employ new dolphin-saving equipment or procedures or to avoid dolphin-associated fishing altogether.

FISHING GEAR

The process of purse seining any species of fish involves the encircling of the school with a long net to form a circular wall of netting. The net must be deep enough to discourage escape underneath it, and the encircling must be done rapidly enough to prevent escape before the ends are closed. The tuna purse seine, described in detail in Chapter 3, is rectangular, typically much longer than it is deep. A seine is approximately 1 mile long and 600 feet deep.

Once the school is located, a skiff is released from the stern of the vessel, with one end of the net (known as the ortza) attached. The skiff anchors this end of the net while the seiner encircles the targeted school and rejoins the skiff. The ortza is transferred to the vessel and made fast, thus closing the circle. At this point the net forms a vertical cylinder around the school of fish. To allow closure of the bottom of the seine, a series of rings are attached to the leadline through which a purse line is run. During the pursing operation, this purse line is pulled in from both ends, choking off the bottom of the seine. When the seine is completely pursed and the rings are alongside the vessel, the process of hauling in, or "drying up," the net can begin.

The remainder of the normal purse-seining operation involves "sacking up" the catch. This consists of reducing the volume of water inside the net until it is possible to bring the catch aboard using a large dip net called a "brailer." This is done through a process of bringing most of the net aboard, leaving only a small sack of reinforced netting in the water to confine the catch for brailing. Once the fish are removed, the remainder of the seine is brought aboard and made ready for the next set.

In fishing for tuna associated with dolphins, a variety of techniques and modifications of the seine are used to minimize the number of dolphins killed (see Chapters 3 and 7).

BEHAVIOR OF TUNA AND DOLPHINS IN THE ETP

Several species of dolphins are found in association with tuna. The spotted dolphin *(Stenella attenuata)* is by far the most important from the point of view of its frequency of association with tuna and its use by fishermen for catching tuna. Three stocks of this species are in the ETP. The frequent appearance of spinner dolphins *(Stenella longirostris)* in sets makes that species significant as well, although in almost all cases it appears in mixed herds with the spotted dolphin. The common dolphin *(Delphinus delphis)* is another important species, although sets on that species are less frequent than on the other two. A few other species found in association with tuna, but much less frequently, are the striped dolphin *(Stenella coeruleoalba)*, the roughtoothed dolphin *(Steno bredanensis)*, the bottlenose dolphin *(Tursiops truncatus)*, and Fraser's dolphin *(Lagenodelphis hosei)*.

The bond linking tuna and spotted dolphins is remarkably strong. It may persist through much or all of the seining operation. During seining, tuna and dolphins continue to associate so tightly that to catch dolphins also means to catch tuna.

FACTORS AFFECTING DOLPHIN MORTALITY

Many factors other than the total fishing effort on dolphins influence dolphin mortality. Some of the most important are the number of tuna caught, vessel captain, species or stock caught, area, flag of vessel, time of capture (day or night), duration of set, presence or absence of strong currents, occurrence of malfunctions, alignment of fine-mesh panel, and use of dolphin-saving procedures.

Many of these factors are interrelated. For instance, the larger the catch of tuna in a set, the longer the set will last, the larger the number of dolphins that will be caught, and the more likely it is that the set will finish after dark (dolphin mortality increases markedly after dark) or that a malfunction will occur. Also, different species or stocks of dolphins have different herd sizes and behaviors and inhabit different areas. Some areas and some species or stocks are fished during only part of the year, so spatial, temporal, and species effects cannot be discriminated. This makes it difficult to identify precisely the cause-effect mechanisms that result in the higher or lower mortalities.

Finally, the skill of the captain has a large effect. Even the best captains occasionally experience a set with high mortality, but in the long run they kill many fewer dolphins than captains of less experience and skill.

ESTIMATES OF DOLPHIN ABUNDANCE

Estimates of dolphin abundance in the ETP have been made by NMFS and IATTC on the basis of observations made from either research vessels or fishing boats. Other methods of estimating abundance, such as mark-recapture experiments, or other sources of data (e.g., sightings from aerial surveys) have proved inadequate for this purpose.

The best available estimate of the average total population of common, spinner, striped, and offshore spotted dolphins in the ETP in 1986–1990 is slightly over 8,000,000. Estimates of absolute abundance are important for seeing the impact of mortality within a stock due to fishing as a proportion of the total population, not as a number without any frame of reference.

The NMFS and IATTC studies demonstrate that none of the indicators of stock size shows any statistically significant trend in the last 5 years. Before 1982, and especially in the late 1970s, several stocks experienced large declines; however, since 1983 all stocks have been stable, and some appear to be increasing. However, the committee notes that better knowledge of recruitment rates and migration patterns of dolphins and better stock identification of individuals are needed for accurate descriptions of population trends.

TECHNIQUES FOR REDUCING DOLPHIN MORTALITY

After extensive analysis, the committee was unable to identify any currently available alternative to setting nets on dolphins that is as efficient as dolphin seining for catching large yellowfin tuna. The committee also could not identify experimental modifications to gear or techniques of catching dolphin-associated tuna that would reduce dolphin mortality to or near to zero and would be practical in the fishery in the immediate future. Therefore, the committee concentrated on incremental improvements and longer-term research and regulatory options (Chapter 7).

Several small modifications to the current methods of tuna purse seining (e.g., the Medina double corkline, jet boats, and the Doppler current profiler) have immediate potential for reducing dolphin mortality. Each of these changes could have an incremental effect. The cumulative effect of these and other innovations could reduce significantly the impact of purse seines on dolphins.

The use of more sweeping changes in purse-seine gear and methods was given major attention by the committee. Modifications to reduce two fundamental problems—canopies and roll-ups—are described in Chapter 7.

The committee believes that the most promising major alterations in purse-seine gear are the following:

- Modifications in netting material.
- Modifications in hang-in ratio.
- Modifications to the purse cable.
- Development of lifting surfaces in critical parts of the net.

Several other concepts explored by the committee also show promise for reducing dolphin mortality. These modifications include inserting barriers between tuna and dolphins, improving the escape of dolphins from the backdown channel, separating the tuna from the dolphins once in the net, releasing dolphins without the backdown procedure, releasing dolphins before backdown, and breaking the tuna-dolphin bond before setting the net.

The committee also explored alternative methods of locating and catching yellowfin tuna when they are not associated with dolphins. These include acoustical methods and fish-aggregating devices.

A variety of techniques are used worldwide to exploit tuna and other midwater schooling fish, but their production rates are much lower than those for the modern purse seiner; therefore, they are not as effective as fishing on dolphins in the ETP for a fishery supplying fish to canneries. Government and industry workshops have explored these alternatives. Methods considered by the committee include live-bait pole-and-line fishing, longlining, midwater trawling, pair trawling, and gillnetting.

Finally, the committee explored regulatory alternatives for reducing dolphin mortality in the ETP purse-seine fishery. The first set of options centers on regulatory alternatives that would further prohibit, directly or indirectly, dolphin mortality. These alternatives include prohibition of dolphin mortality and issuance of dolphin-mortality certificates.

The second set of options centers on alternatives that would create incentives for behavior that reduces dolphin mortality, as opposed to direct or indirect prohibition on dolphin mortality itself. These include incentives for tuna fishing with alternative gear, price incentives for fishing on non-dolphin-associated tuna, and the development of vessel-captain performance standards coordinated with a training and evaluation program.

RECOMMENDATIONS

The committee's recommendations are presented in two parts. The first part treats avenues for developing promising new techniques for reducing dolphin mortality in the existing purse-seine fishery on dolphins. The second part treats research on and development of new methods of harvesting ETP yellowfin not in association with dolphins.

Recommendations Concerning the ETP Tuna-Dolphin Fishery

The committee judges that improvement in captain performance is the single most important step that can be taken to reduce dolphin mortality in the ETP purse-seine fishery. *Therefore, the committee recommends that an international meeting be convened of representatives of government and industry from all countries engaged in the ETP purse-seine fishery.* The purposes of this meeting would be the following:

1. Develop an educational certification and monitoring protocol for all captains in the international fleet.
2. Initiate research on the development of incentives to improve captain performance.

The committee recommends that two approaches, short term and long term, be undertaken in gear and methods research and development. These approaches are the following:

1. A number of small modifications of current methods could be built and tested immediately on commercial fishing trips. The most promising modifications are the current profiler, jet boat, double corkline, pear-shaped snap rings, and polyester net. The committee emphasizes that it is of paramount importance that an international program be developed to systematically deploy, test, and evaluate these modifications of current methods.
2. A number of major modifications of current methods need to be researched and developed on a long-term basis. These modifications include inflatable sections or partitions in the net, lifting surfaces, modified purse cable, new netting materials, and modified net designs. The committee recommends a long-term engineering approach toward eliminating major causes of dolphin mortality in the purse-seine process—canopies, roll-ups, and collapses in the backdown channel.

Research is needed for a better understanding of the behavior of both dolphins and tuna and the bond between them. Details, costs, and potential benefits of many of the concepts delineated above cannot be judged at present.

Fishermen need both incentives and options to make further progress in reducing dolphin mortality. The individual fisherman alone cannot be expected to develop the options that offer significant improvements if they represent major changes to the present gear because fishermen who adopt gear that is likely to reduce dolphin mortality might incur economic losses. *The committee recommends that a program of research be established to develop options and to demonstrate them to the industry.* Such a program should include two facets:

1. An experimental research program of innovative gear to investigate performance and techniques. This program would have access to a modern commercial purse seiner as a dedicated vessel that would not be constrained by the normal pressures of tuna productivity. Because the capture of animals by fishing gear involves interactions between the animals and the gear, the research must include a program of behavioral studies focused on the reactions of both tuna and dolphins to fishing gear and other stimuli. Techniques would include underwater video observation by remote-operated vehicle and acoustic sensing and tracking. Most of the effort would be at sea with deployed gear.

2. The information gained from the research above would then be used to develop rational purse-seine modifications and alternative harvesting methods based on the engineering requirements of the fishery. This focus would use analysis and modeling to develop, refine, and evaluate each concept to ensure a reasonable chance of success before it was attempted as a commercial prototype. After the analysis and modeling, promising prototypes would be field-tested.

The committee recommends that a program of research on the behavior of tuna and dolphins be established and that it have the following components:

1. Oceanographic correlates.
2. Simultaneous tracking of dolphins and associated yellowfin.
3. Tracking flotsam-associated yellowfin.
4. Studies using fish-aggregating devices (FADs).
5. Satellite monitoring of radio-tagged dolphins.

Recommendations for Research on Harvesting Tuna Not in Association with Dolphins

Even if harvest of yellowfin tuna in association with dolphins continues, the committee recommends that major research be undertaken to explore new methods of harvesting yellowfin not in association with dolphins in the ETP. The promising avenues of research identified by the committee are as follows:

1. Research into the night behavior of tuna and dolphins. If large yellowfin

do not associate with dolphins at night, purse seining or trawling could be done at night, which would reduce dolphin mortality significantly.

2. Research into new methods of purse seining.

3. Research on existing FADs and new technologies such as submerged FADs, which may have a greater potential than surface FADs for attracting and holding commercially harvestable schools of large yellowfin tuna.

4. The use of satellite oceanographic techniques to locate aggregations of tuna not associated with dolphins in the ETP.

5. The use of alternative techniques of locating tuna that do not depend on the sighting of dolphins.

6. Assessment of the impact on tuna populations if purse seining for tuna in association with dolphins is discontinued.

1

Introduction

Approximately one quarter of the world's tuna catch takes place in the eastern tropical Pacific Ocean (ETP). In that area (usually defined as east of 150° W, sometimes 160° W), the most economically important tuna species, the yellowfin *(Thunnus albacares)*, often is found in association with various species of dolphins.* Since the late 1940s, tuna fishermen increasingly have taken advantage of this association and catch tuna by setting their nets around the highly visible herds of dolphins, which, being mammals, must surface often to breathe. Despite improvements in techniques and in gear that have substantially reduced the number of dolphins killed in the ETP tuna fishery, thousands of dolphins are still killed each year.

When the Congress reauthorized the Marine Mammal Protection Act of 1972 on November 23, 1988, it stipulated several amendments to the act. One of these amendments (Section 110 (a)) focused on identifying appropriate research into promising new methods of locating and catching yellowfin tuna without the incidental capture of dolphins. It further directed the Secretary of Commerce to "contract for an independent review of information pertaining to such potential alternative methods to be conducted by the National Academy of Sciences with individuals having scientific, technical, or other

---

* Dolphins and porpoises are small, toothed, whale-like marine mammals (cetaceans). The term "dolphin" is usually applied to members of the family Delphinidae; "porpoise" usually refers to members of the family Phocoenidae. Some people refer to the whole group as porpoises; others, as dolphins. Some confusion results from the latter term because the fish *Coryphaena hippurus* is also commonly called dolphin (this is the American Fisheries Society's approved common name (Robins et al., 1991)). However, because the cetaceans most affected by the tuna fishery are members of the family Delphinidae, we refer to them as dolphins in this report.

expertise that may be relevant to the identification of promising alternative fishing techniques." The report from the National Academy of Sciences would then be submitted by the Secretary, together with a proposed plan of research, development, and implementation of alternative fishing techniques, to the Committee on Commerce, Science, and Transportation of the Senate and the Committee on Merchant Marine and Fisheries of the House of Representatives.

This report provides background information on catching yellowfin tuna in the ETP and the incidental capture of dolphins, recommendations from workshops focusing on improving current fishing techniques, and recommendations from the Committee on Reducing Porpoise Mortality from Tuna Fishing for research on alternative techniques.

THE PROBLEM

Catching tuna with purse seines is the most productive method available. There are three major modes of fishing this way. The first is school fishing, in which schools of tuna near the surface are found visually and the purse seine is set around the school. The second, called log fishing, depends on the attractive power of floating objects for tuna. Purse seines are set around natural logs or fish-aggregating devices and catch the fish that are often associated with them. The third method—the focus of this report—is "dolphin fishing." Yellowfin tuna and certain species of dolphins are often associated, especially in the ETP. (The extent and effects of dolphin-associated tuna fishing in other oceans are poorly quantified or unknown.) Dolphins frequently come to the surface to breathe and thus are easy to see from a boat. The purse seine is set around the dolphin herds, and because the tuna are closely associated with them, catching dolphins usually means catching tuna as well. Although a variety of techniques and equipment have been developed to release the dolphins safely, thousands are still killed each year by dolphin fishing.

Although all three methods of purse seining catch tuna, log and school fishing catch mostly small, sexually immature tuna. Dolphin fishing usually catches large fish that are often sexually mature and produces larger average catches than the other methods. Thus, redirecting the fishing away from tuna associated with dolphins would be less efficient and would have a negative effect on the yield of the fishery and perhaps on the conservation of tuna populations. From the point of view only of economics and harvesting tuna, large tuna should be sought; fishermen should fish on dolphins and be discouraged from fishing on logs or schools. However, dolphin fishing kills dolphins; to minimize the killing of dolphins, all fishing should be directed away from dolphins. This dichotomy is the basis of the tuna-dolphin problem.

Two operational factors are related to the total number of dolphins killed in purse seining for tuna. The first is the number of times purse seines are set around dolphins, and the second is the average number of dolphins killed in each such set. The first factor is affected by market prices and availability of tuna of different sizes, restrictions such as the policy of some processors not to buy tuna caught on trips that have involved the intentional encirclement or death of dolphins (so-called "dolphin-safe" tuna), and the availability of alternative methods of catching tuna. The second factor depends on conditions such as the fishermen's motivation, skill, and experience; the condition of the vessel's equipment; weather conditions; and technological developments. Therefore, two general approaches are available to achieve the goal of reducing dolphin mortality—economic and technical—and each approach has several options.

BACKGROUND ON REDUCING DOLPHIN MORTALITY FROM TUNA FISHING

The Current Tuna Fishery and Its Development

More than 70 nations participate in the world tuna fishery, but only 10 of them account for almost 85% of the catch. They are Japan, which accounts for about 29%; the United States, 12%; Spain, 8%; the Republic of Korea, 6%; the Philippines, France, Indonesia, and Mexico, 5% each; the Republic of China, 4%; and Venezuela, 3% (IATTC, 1989a). About 36% of this catch is consumed by Japan and about 31% by the United States (IATTC, 1989a). Over the years, one of the most productive tuna fishing areas has been the ETP. On an average, this area accounts for nearly one quarter of the world tuna catch (IATTC, 1989a).

Albacore or longfin tuna *(Thunnus alalunga)* has been fished commercially off southern California since the turn of the century. It was the first tuna to be canned and is the only tuna known as "white-meat" tuna. As the demand for canned tuna increased, southern California fishermen ventured off the coast of Mexico to catch yellowfin and skipjack tuna *(Katsuwonus pelamis)* and by the 1920s, by which time bait could be kept alive for days, boats ventured farther offshore. In the 1930s, mechanical coolers made it possible to freeze the catch on board, further freeing the vessels to explore offshore waters; a fishing area developed as far south as the equator and up to several hundred miles offshore where yellowfin and skipjack tuna were the target species. Synthetic webbing and the advent of the Puretic power block in 1955 led to the completion of transforming bait boats to ocean-going purse seiners capable of staying at sea for a month or more. The fishery continued to expand southward to Peru and Chile, and during the late 1960s and the 1970s, it expanded farther offshore to about 145° W. The capacity to operate far offshore led vessels to areas where yellowfin tuna were more abundant and where they associated more frequently with dolphins than they did nearer the coast (Hofman, 1981). A major fishery for yellowfin caught in association with dolphins developed. In this fishery, purse seines are set around herds of dolphins and schools of tuna with the intention of retaining the tuna and releasing the dolphins. (Skipjack tend to be caught close to shore, and less than 0.4% of the total catch is caught in association with dolphins.)

These changes and economic considerations made purse seining so efficient that by 1961 nearly all of the larger bait boats had been converted to purse seiners. Since then, purse seiners have been the dominant type of vessel in the fishery (98% of the mean 1984–1988 capacity) (see Table 2 in IATTC, 1989a). The annual catch of yellowfin tuna in the Pacific Ocean east of 150° W was approximately 159,000 metric tons in 1980 with a value of approximately $192 million. After a decline in the El Niño years of 1982–1983, the catch in the area increased to 269,000 metric tons in 1986 (IATTC, 1987).

In Table 1-1, the Inter-American Tropical Tuna Commission (IATTC) data on the fishery for yellowfin in the ETP are broken down by fishing mode (dolphin vs. non-dolphin), area (inside and outside 200-mile national Exclusive Economic Zones), and fleet (U.S. vs. non-U.S.). Dolphin fishing includes all sets made on tuna associated with dolphins; non-dolphin includes all other purse-seine sets. Note that (1) 67% of the dolphin catch is taken outside the Exclusive Economic Zones, and 84% of the non-dolphin catch within them; (2) 77% of the U.S. catch of yellowfin in the ETP and 59% of the non-U.S. catch are dolphin catch; and (3) 84% of dolphin sets involve spotted dolphins, in either pure or mixed herds. In Figures 1-1 and 1-2 it can be seen that the

TABLE 1-1A　Mean ETP Yellowfin Catch in Thousands of Metric Tons, by Fishing Mode, Area, and Fleet, 1984–1988[a]

Area and Fleet	Dolphin	Non-dolphin	Total
Inside 200 miles	50.8	73.5	124.3
Outside 200 miles	101.4	12.0	113.3
Total	152.2	85.5	237.6
U.S. Fleet	66.1	19.4	85.5
Non-U.S. Fleet	89.2	62.4	151.6
Total	155.3	81.8	237.1

TABLE 1-1B　Percentage of Sets on Different Types of Dolphin Herds, 1984–1988[a]

Type of Dolphin Herd	Percentage
Spotted Pure	45.5
Spotted + Eastern Spinner	21.9
Spotted + Whitebelly Spinner	16.5
Spinner Pure	1.7
Common Dolphin	4.6
Other	9.8
Total	100.0

[a]Data from IATTC.

geographic distribution of the catch of yellowfin and skipjack tuna is different for dolphin and non-dolphin fishing. The figures refer to April–September 1986, but are representative: in general, tuna associated with dolphins are caught farther offshore than those caught in non-dolphin fishing.

Until recently, the fishery was dominated by the U.S. fleet. A small tuna fishery existed in Mexico before World War II, and during the 1950s, tuna fisheries were developed in Peru and Ecuador. However, the fleets were small and the combined catch of these countries at that time amounted to less than 10% of the total catch from the eastern Pacific. Since 1960, fleets of nations other than the United States have increased greatly by construction of new vessels and transfer of flags, and the U.S. fleet has decreased. The proportion of the catch taken by U.S. vessels decreased from 90% in 1960 to 32% in 1988 and 11% in 1991, and that taken by Latin American countries bordering the Pacific Ocean increased from 10% to 47% in 1988 and 57% in 1991 (IATTC, 1989a; M. Hall, IATTC, La Jolla, Calif., personal commun., 1991).

Until about 1975, the U.S. market consumed about 80% of the catch of surface-caught (i.e., by bait-poles and purse seines) tuna from the ETP. With

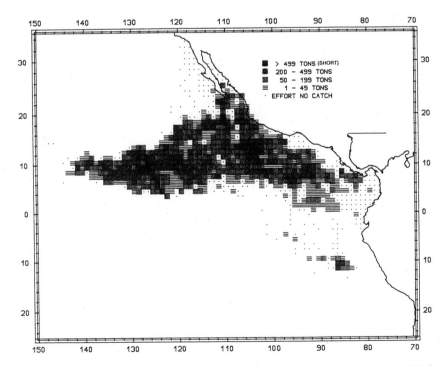

FIGURE 1-1 Dolphin-associated sets. Eastern Pacific real distribution of yellowfin and skipjack tuna purse-seine catch for April–September 1986 from usable log-book data. (1 short ton = 0.907 metric ton.) Source: Richard Deriso, IATTC, La Jolla, CA, unpublished material.

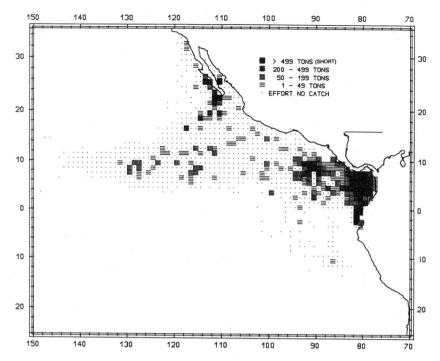

FIGURE 1-2 Non-dolphin-associated sets. Eastern Pacific areal distribution of yellowfin and skipjack tuna purse-seine catch for April–September 1986 from usable log-book data. (1 short ton = 0.907 metric ton.) Source: Richard Deriso, IATTC, La Jolla, CA, unpublished material.

the expansion of the non-U.S. fleets during the 1980s, the share of the catch absorbed by the U.S. market declined, falling to 45% in 1987. Most of the difference went to Latin America, Europe, and the Far East (IATTC, 1989a).

Of particular interest to the committee is the surface fishery for yellowfin tuna in the ETP. Surface-caught yellowfin and skipjack tuna provide the raw material for the canned light-meat tuna product marketed primarily in the United States, Europe, and the Far East. Over the years, as new fishing areas have become viable, the ETP has contributed a smaller fraction of the world catch of yellowfin and skipjack.

Dolphin Mortality Associated with the Tuna Fishery

Although no accurate data are available for the early years (1950–1972), the increased offshore operation of increasingly sophisticated fishing vessels setting their nets on dolphin herds clearly led to very high mortality in the ETP. Estimates of dolphin mortality (e.g., Perrin, 1968, 1969; NMFS, 1980a; Lo and Smith, 1986) are based on extremely small, nonrepresentative, and probably biased data sets, vary from one investigator to another, and vary from year to year. Nonetheless, it seems likely that from 1960 to 1972, more than 100,000 dolphins were killed each year by the U.S. fleet. After 1972, the annual number killed by the U.S. fleet declined to approximately 20,000 in 1979—23,595 according to Hofman (1981) and 17,938 according to the National Marine Fisheries Service (NMFS, 1980a); IATTC (1991a) estimated total 1979 mortality at 21,467. In 1988–1990, the annual U.S. kill according to NMFS was 19,712 in 1988, 12,643 in 1989, and 5,083 in 1990 (MMC, 1991). This decline was due to a decrease in the rate at which tuna boats killed dolphins (Wahlen, 1986) and a decrease in the number of boats in the U.S. fleet. Most of the tuna fleet is now registered in other countries; by 1986, only 34 boats of the 103-boat fleet were registered in the United States and subject to NMFS regulations (NMFS, 1987). IATTC (1991a) estimated that the total kill for 1989 was 97,000–102,000 dolphins, of which the 12,643 attributed to the U.S. fleet was less than 15%. In 1990, the total mortality declined to 52,000–56,000 dolphins (IATTC, unpublished data), less than 10% of which (5,083) was caused by the U.S. fleet. To date in 1991 (M. Hall, personal commun., 1991), average mortality per set has fallen to half of the 1990 values; this, added to a reduction of fishing effort on dolphins, is projected to result in a total mortality of about 25,000 dolphins for the whole international fleet in 1991.

Three species of dolphins are most commonly affected: the spotted dolphin *(Stenella attenuata)*, the spinner dolphin *(Stenella longirostris)*, and the common dolphin *(Delphinus delphis)*. NMFS, which has responsibility for protecting marine mammals under the Marine Mammal Protection Act of 1972, set an annual quota in 1980 of 20,500 dolphins that could be killed by U.S. tuna boats, that quota including subquotas for various species and stocks. Additional regulations apply to fishing techniques.

Reducing Dolphin Mortality

Improvements in fishing gear and techniques discussed in Chapters 3 and 7, the mandatory observer program, and the kill quota were responsible for the dramatic decline in the number of dolphins killed by boats in the U.S. fleet. Currently, only about 0.5% of dolphins encircled by purse seines are injured or killed and almost all of those are killed in a small number of "problem" sets (NMFS, unpublished data). Most sets on dolphins are zero-kill sets.

Three major advances have influenced the reduction in dolphin kills. The first is a procedure developed by tuna fisherman Anton Misetich known as "backdown," in which the seine is pulled out from under the dolphins, thus allowing them to escape without losing the fish in the net.

The second advance is the use of Medina panels. Designed by tuna fisherman Harold Medina, strips (panels) of relatively fine-mesh sections of net about 33 feet deep are placed adjacent to the backdown area and below the corkline. The fine mesh is small enough to prevent the entanglement of the

the corkline. The fine mesh is small enough to prevent the entanglement of the snouts and flippers of dolphins in the net and thus has a major effect in reducing dolphin mortality.

Third, several methods have been developed to aid the release of dolphins from the net, such as the use of speedboats equipped with towing bridles to keep the net open and the use of a raft inside the net to facilitate hand rescue.

Many other modifications of gear and techniques have been tried, but the three mentioned above have played the largest role in reducing mortality. The NMFS regulations reflect this: They require all purse seiners to use Medina panels and to carry at least two speedboats and an inflatable raft. Backdown is required. Night sets on dolphins are also prohibited and bringing live marine mammals on board is prohibited.

All those modifications were implemented more than a decade ago. More recent declines in mortality can be traced to improvements in equipment and the performance of the fishermen. Industrial groups have set up national technical advisory offices that in collaboration with IATTC include organizing training courses for vessel captains and crews, inspecting the condition and performance of dolphin-saving gear, and diagnosing individual vessel or skipper problems. The Porpoise Rescue Foundation, funded by the tuna industry, has also contributed ideas and methods to reduce dolphin mortality. The recent declines in mortality, from 130,000 dolphins in 1986 to perhaps as few as 25,000 in 1991, were achieved without any major technological advances or additions to the fishing gear. Although mortality of dolphins has been drastically reduced, the annual kill is still substantial. In addition to being of humanitarian and ecological concern, the killing of dolphins is detrimental to fishermen. For example, they have to disentangle the animals from the nets, a long and sometimes dangerous procedure. Furthermore, fishermen view the dolphins as a valuable resource because they attract and hold large yellowfin tuna near the surface where they can be caught. Thus, reducing dolphin mortality as much as possible is in their interest.

Several workshops have focused on improving current fishing techniques and the development of alternative ones (Ralston, 1977; Hofman, 1979, 1981; NMFS, 1986; DeMaster, 1989). Participants in these workshops included representatives of the U.S. Navy, the National Science Foundation, the tuna and other industries, the Porpoise Rescue Foundation, academe, the Marine Mammal Commission, NMFS, IATTC, and various environmental organizations. The workshops have been distinguished by a spirit of cooperation. At a workshop held in October 1988 several points received consensus.

• No single alternative to purse-seine fishing on dolphins has proved acceptable thus far to the tuna industry.

• The industry will continue to work toward reducing dolphin mortality due to tuna fishing, but if purse-seine fishing on dolphins continues, some dolphin mortality will occur.

• Available information is insufficient to identify or evaluate alternative methods; additional research is necessary and desirable.

• Research should focus on alternative methods that appear most promising. These are using longlines; understanding and breaking the tuna-dolphin bond; locating and catching schools of yellowfin tuna not associated with dolphins; and using new technology such as fish-aggregating devices, long-range sonar, and remote sensing.

THE COMMITTEE'S STUDY

Because of concerns over the incidental capture and death of dolphins in operations of yellowfin tuna fishing, the U.S. Congress requested that the National Research Council (NRC) form a committee to search for promising new techniques for locating and catching yellowfin tuna without killing dolphins. The NRC Board on Biology, in collaboration with the Board on Environmental Studies and Toxicology and in consultation with the Marine Board, has convened this committee of experts on tuna and dolphin biology (including behavior and ecology), fishing-gear design and operation, fishery and wildlife management, resource economics, hydroacoustics and communication systems, and remote sensing. The committee has reviewed aspects of the scientific and technical information concerning the following topics, as they bear on the tuna fishery and the incidental killing of dolphins:

• Biology and ecology of yellowfin tuna.
• Biology, ecology, and behavior of dolphins at risk from purse seining.
• Economic and management implications of harvesting tuna of various sizes and ages.
• Development of economically viable alternatives to the setting of nets on dolphins.
• Nature of the tuna-dolphin bond and methods of breaking the bond while in and around nets.
• Evaluation of potential research leading to cost-effective methods of finding and catching tuna without killing dolphins.

The committee did not evaluate various target levels of either dolphin population size or mortality. Some target levels are determined by ethics alone (e.g., zero mortality) and not science, whereas other target levels can be grounded scientifically (e.g., optimal sustainable population). The committee was not charged with selecting among various methods or targets for reducing dolphin mortality (for a discussion of this topic see Aron, 1988).

2

Some Policy and Economic Considerations

U.S. TUNA AND MARINE MAMMAL POLICY

The U.S. policy and regulatory framework relevant to the tuna and dolphin issue is a subset of general U.S. fisheries and marine mammal policy. In this section, we briefly summarize the U.S. laws and policies that have had significant effects on the tuna-dolphin issue. We first describe the general policy framework for tuna fisheries and then the marine mammal policies that have affected the tuna industry.

U.S. Tuna Policy

The U.S. government has been involved in some form of tuna-policy negotiations since the 1940s and in formal regulatory activity concerning tuna fisheries since the 1960s. These laws and activities include the founding of IATTC in 1949 and the Tuna Conventions Act of 1950; the Fishermen's Protective Act (1954); the International Commission on the Conservation of Atlantic Tunas and the Atlantic Tuna Conventions Act (1975); the U.S.-Canada Pacific Albacore Treaty (1982); the South Pacific Tuna Act (1988); and others (Orbach and Maiolo, 1989). IATTC and the Tuna Conventions Act and the Fishery Conservation and Management Act of 1976 (FCMA) have had the greatest impact on the tuna industry. The Tariff Act of 1930 (and amendments) and the 1946 General Agreement on Tariffs and Trade (GATT) have also affected the industry.

IATTC and Tuna Conventions Act

Although there have been recent significant shifts in fishing effort by the U.S. high-seas tuna fleet to areas such as the western Pacific, the majority of the fleet's activity in the last 45 years has been in the ETP (see Figure 2-1 for recent distribution of catch in the ETP) off the coasts of countries from Mexico to Peru. In 1949, IATTC was formed by treaty between the United States and Costa Rica; a number of other coastal or fishing nations have subsequently adhered to the convention. The commission was codified into U.S. law through the Tuna Conventions Act of 1950. The original purpose of IATTC was to address and research issues pertinent to the conservation of tunas. For a number of reasons the attention of the commission has since expanded to include other large pelagic species such as billfish and coordination of certain international marine mammal conservation efforts (Joseph and Greenough, 1979).

Beginning in the early 1960s, IATTC recommended a series of yellowfin tuna quotas to be implemented through seasonal closures in the commission's

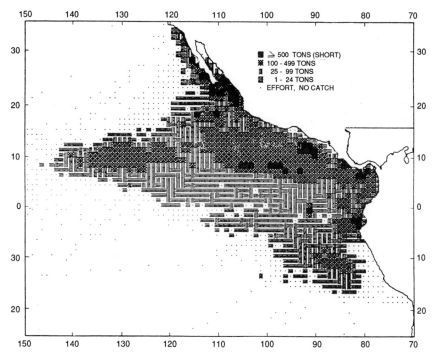

FIGURE 2-1 Average annual catches of yellowfin in the eastern Pacific during 1979–1987 for all purse-seine trips from usable log-book data. (1 short ton = 0.907 metric ton.) Source: Adapted from IATTC, 1989b.

Yellowfin Regulatory Area of the ETP (see Figure 2-2). The United States was the only member nation of IATTC that did not claim jurisdiction over tuna within its 200-mile resource-control zone under its national fisheries legislation (this position has recently changed, as mentioned below). All other nations claimed such jurisdiction (see below). This difference in national fisheries policy, along with competition for the tuna resource itself, led to friction among the member nations that included seizures of U.S. tuna vessels by other member nations, retaliatory embargoes by the United States of tuna products from other member nations, and disagreements over national allocations. This situation led to the withdrawal of Mexico from the commission in 1978 and Costa Rica in 1979 (Van Dyke and Heftel, 1981). Costa Rica has recently rejoined the commission. Vanuatu, a nation new to the fishery, has also joined, and Colombia and Venezuela have applied for membership. Current members of IATTC are Costa Rica, France, Japan, Nicaragua, Panama, the United States, and Vanuatu.

IATTC has continued its activities, however, in research and data collection for ETP tuna fisheries and has developed a principal role in the international marine mammal conservation effort, in particular in research efforts concerning the tuna-dolphin issue in the ETP. IATTC has been successful in obtaining the cooperation of both member and nonmember nations in data collection concerning dolphin mortality and, in many cases, in facilitating the cooperation of these nations in developing marine mammal protection regulations. However, no tuna management regime has existed in the ETP since 1979 (Orbach and Maiolo, 1989). All nations fishing for tuna in the ETP participate in the IATTC tuna-dolphin program, which includes the observer program, programs to reduce dolphin mortality through diagnosis and solution of gear problems, and training programs for captains and crews.

Fishery Conservation and Management Act

The FCMA is the principal U.S. federal law governing the management of domestic U.S. fisheries. Under the FCMA, until its amendment in 1990 the United States claimed jurisdiction over all fish species within 200 miles of its coast except "highly migratory species." The act defines these as tuna "which, over the course of their life cycle, spawn and migrate over great distances in waters of the ocean" (Section 3). This provision, the so-called "tuna exemption," was lobbied for by the U.S. high-seas tuna fleet. Their principal argument was that the U.S. fleet must fish for tuna off the shores of other nations, often within the 200-mile jurisdictions of these countries, and because all other nations include tuna in their jurisdictions, the U.S. fleet needed this provision to claim exemption from the laws of those nations. Under this law and the Fisherman's Protective Act, the United States has implemented its embargoes and other reactions to the seizure of U.S. vessels by Latin

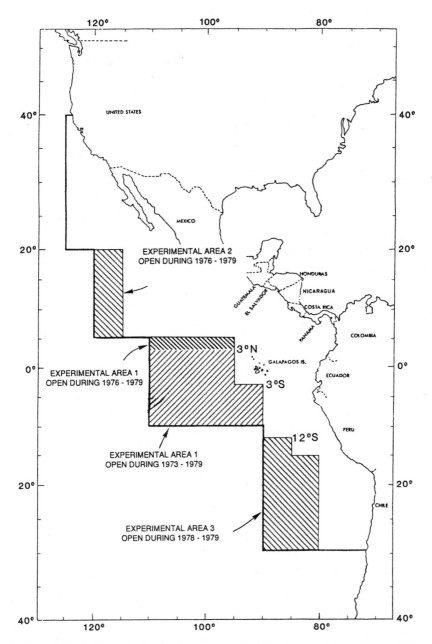

FIGURE 2-2 IATTC's Yellowfin Regulatory Area (CYRA). Source: Adapted from IATTC, 1989b.

American nations. The FCMA was amended in 1990, and the United States now includes tuna in its jurisdiction.

Although the circumstances surrounding U.S. tuna policy under the FCMA are extremely complex, two important implications of this policy are important for the tuna-dolphin issue. (1) The U.S. tuna exemption, among other factors, has strained the relationship between the United States and its Latin American neighbors to such an extent that it is expensive and sometimes difficult for U.S. vessels to fish inside the 200-mile resource-control zones of many of these countries. Thus, the U.S. fleet must fish primarily outside the 200-mile zones where the incidence of the tuna-dolphin relationship is greater. (2) Much of the potential for international cooperation on marine mammal issues is dependent on the general marine policy relationship among the countries bordering the ETP, which has been adversely affected by the U.S. tuna policy (Cicin-Sain et al., 1986). Recent amendments to the FCMA may alleviate these difficulties.

U.S. Marine Mammal Policy

In addition to the general U.S. tuna-policy framework, the U.S. policies on marine mammal conservation and protection have significantly affected the U.S. high-seas tuna fleet. The most significant of these policies result from the Marine Mammal Protection Act of 1972 (MMPA).

Marine Mammal Protection Act

The MMPA directs that all marine mammal populations be managed for their "optimum sustainable population," defined as "the number of animals which will result in the maximum productivity of the population or species" (Section 3). This law took effect in December 1972, at which time it placed a moratorium on the "taking" of marine mammals; "taking" is defined as "to harass, hunt, capture or kill ... any marine mammal" (Section 3). Two exceptions were made to the moratorium. The first exception was taking marine mammals for subsistence or traditional native handicraft purposes by native Americans, and the second was taking marine mammals in the course of commercial fishing operations. These exceptions could not be granted when the marine mammal population in question was endangered. Special conditions were required even when the population was not endangered. For example, in issuing permits for taking marine mammals during commercial fishing operations, the Secretary of Commerce was required to consider, among other things, "the conservation, development, and utilization of fishery resources" and "the economic and technological feasibility of implementation" (Section 103). The MMPA (Section 1371) was amended in 1981, to state that " ... [the] goal [of zero mortality] shall be satisfied in the case of the incidental taking of marine mammals in the course of purse-seine fishing for yellowfin tuna by a continuation of the application of the best marine mammal safety techniques and equipment that are economically and technologically practicable."

The MMPA is important to the U.S. high-seas tuna fleet because setting of nets on yellowfin tuna in association with dolphins has traditionally resulted in significant dolphin mortality. Beginning in the early 1970s, regulations began to be placed on the U.S. high-seas fleet under the MMPA. In 1975, NMFS, the branch of the U.S. Department of Commerce with delegated authority for marine mammal regulations under the MMPA, instituted a quota of 78,000 dolphins as the maximum that could be "taken" by the U.S. tuna fleet per year. Accompanying this quota, which was designed to be reduced each year until it reached "insignificant levels," several gear and procedural restrictions were also placed on U.S. vessels. Significant among these were requirements for the use of the Medina panel and the backdown procedure (Coe et al., 1984). In 1981, the quota for dolphin mortality was set at 20,500 animals per year, a mortality level that had been achieved by the U.S. fleet since 1977, shortly after the initial regulations were put into place.

In 1988, the MMPA was reauthorized and amended. Additional regulations have been developed on the basis of these amendments. Full observer coverage on all U.S. high-seas tuna vessels is required, and a prohibition has been placed on sets after sunset and on the use of certain explosives in sets on yellowfin associated with dolphin. Restrictions were also placed on the percentages of certain species and stocks in the total dolphin mortality.

Perhaps the most important new regulations were those concerning the U.S. embargoes of tuna products from countries that do not have marine mammal protection regulations comparable to U.S. regulations or whose performance with regard to dolphin mortality exceeds the standard set by the U.S. fleet. These regulations stipulate that all countries whose vessels seine yellowfin tuna in association with dolphins must achieve a dolphin mortality rate of no more than two times the U.S. fleet mortality rate by 1989, and no more than one and one-fourth times the U.S. fleet rate by 1990. If these rates were not achieved, the embargo provisions would be activated. (An embargo on Mexican tuna was imposed in October 1990, and shortly afterward tuna from Vanuatu and Venezuela were also embargoed. A panel of GATT recently (1991) ruled that the embargo violates the provisions of GATT.)

International Policy Context

The tuna fishery is part of a highly capitalized and international food-processing industry. This situation, combined with the migratory nature of the tuna, results in complex policy and management conditions that are critical to any effort to preserve and conserve dolphin populations (Joseph and Greenough, 1979).

At present, the most critical element in this international policy is the relationship between the United States and the other countries fishing for tuna in the ETP. The United States has the strictest set of marine mammal protection laws and policies among the ETP fishing nations. These laws include 100% observer coverage on all U.S. tuna vessels. The most significant organization assisting the other ETP fishing nations in the dolphin monitoring and protection effort has been IATTC, which runs an observer program in which ETP fishing nations cooperate on a voluntary basis. In this program, currently 100% of trips by boats from Vanuatu, Ecuador, Panama, Venezuela, and the United States carry observers. Other countries carry observers on approximately one-third of their trips. Overall, close to 57% of all 1991 departures for non-U.S. trips carry observers. It is through this program that data are obtained concerning the levels of dolphin mortality in the foreign fleets.

As noted elsewhere in this report, the relationship between the United States and many of the other ETP fishing nations—notably Mexico—concerning fishery resources has not been smooth. With Mexico in particular, the arrest and confiscation of U.S. vessels in Mexican waters and U.S. embargoes against Mexican tuna products have created considerable tension between the two countries (Cicin-Sain et al., 1986).

This circumstance is important because the 1988 amendments to the MMPA provide for new embargoes against the tuna products of any nation whose fleet dolphin-mortality rates do not meet the U.S. standards noted above. The recent embargo on Mexican tuna and the subsequent GATT decision favoring Mexico have complicated matters and made it even more difficult to predict how relationships between the U.S. and embargoed countries will affect their cooperation in the voluntary IATTC observer program and the ability to monitor or enforce restrictions on dolphin mortality in the non-U.S. portion of the ETP fleet.

The potential embargoes under the MMPA, the actions of the tuna-processing firms, and the responses of the U.S. and international fleets to these events will be critical in determining the ability to monitor and enforce any prohibitions on dolphin mortality associated with tuna fishing.

ECONOMIC CONSIDERATIONS

Apart from the present backdown procedure, the Medina panels, and improved techniques for releasing dolphins from nets (including improved net design and speeding up of the process), most technological innovations that have been tested over the years have failed for one reason or another to reduce the incidental kill of dolphins during tuna purse seining. It is clear that, as currently practiced, purse seining for tuna associated with dolphins will continue to kill some dolphins in the ETP and in other oceans where it is used. Engineering solutions to the problem may exist but experience so far suggests that "promising new techniques for finding and catching yellowfin tuna without killing porpoises" (statement of task for the committee) in the ETP are elusive, may be costly to develop, and may require considerable investments in new vessels and equipment. To put such an overall effort in perspective, it is useful to consider the economic structure of the modern tuna industry and the relative contribution of dolphin-associated fishing to the world tuna harvest and international tuna markets. Such a perspective may also reveal the difficulties of developing national policies and programs to deal with dolphin mortality in this international fishery. Sakagawa (1991) provides a more detailed discussion of these matters.

Recent Tuna-Harvesting Operations

During 1988, the global tuna harvest was 2.5 million metric tons (see Table 2-1). The cumulative maximum sustainable yield from all world tuna fisheries was estimated to be about 3.5 million metric tons (R. Francis, University of Washington, Seattle, personal commun., 1990), so the annual tuna harvest from all oceans could increase if all the world's tuna fisheries were fully exploited. Not all of the world's tuna resources can be exploited commercially using current fishing methods, so the annual harvest from world tuna fisheries may never reach the global maximum sustainable yield. During 1989, approximately 232,000 metric tons of tuna or about 9% of the global tuna harvest was taken in association with dolphins in the ETP (NMFS, unpublished data). All of this dolphin-caught tuna was harvested from the ETP by 123 large high-seas purse-seine vessels from nine nations (IATTC, unpublished data). Twenty-nine of these vessels operated under the U.S. flag; they accounted for 26% of the tuna harvest and about 12% of dolphin mortality during 1989 (NMFS, unpublished data). The remaining 94 vessels accounted for 74% of the tuna harvest and about 88% of the dolphin mortality.

TABLE 2-1 World Tuna Harvest in Thousands of Metric Tons by Ocean, 1981–1988[a]

Ocean	1981	1982	1983	1984	1985	1986	1987	1988
Atlantic	421	470	451	385	439	396	387	391
Pacific	1,208	1,193	1,290	1,440	1,366	1,599	1,593	1,643
Indian	144	199	221	271	319	342	384	473
Total	1,773	1,862	1,961	2,096	2,123	2,338	1,365	2,507

[a]Data from the Food and Agricultural Organization of the United Nations (FAO, 1984, 1989).

The national flag of a high-seas tuna purse-seine vessel is important because it determines under which set of national laws the vessel operates, at which ports it may deliver fish, and the nature of its operating costs. However, the national flag of the vessel reveals very little about the impact of that vessel's operations on any national economy. For example, with an average crew of 17 per vessel, about 500 fishermen were employed aboard the 29 U.S. tuna purse-seine vessels that engaged in dolphin-associated fishing during January 1990, but fewer than one-third of them were U.S. citizens. Since vessels can deliver fish to ports outside their home countries, and they refuel, refurbish supplies, and make repairs primarily at those ports, they often generate economic impacts and "multiplier effects" (King and Bateman, 1985) primarily outside their home countries. On the other hand, a number of U.S. citizens work in the foreign fleets, as skippers, helicopter pilots, navigators, deck bosses, or fleet managers, and may send their earnings back to the United States. These fleets also generate economic benefits for the United States through the involvement of U.S. companies and citizens in activities such as equipping and maintenance of vessels and shipping and handling of fish. Most vessels of all flags use engines, winches, power blocks, helicopters, and other equipment manufactured in the United States. Thus, although few U.S. boats currently participate in the ETP dolphin-associated tuna fishery, that fishery continues to have some economic importance to the United States.

Historical Perspective

Until about 1975, U.S. tuna companies assured themselves of reliable raw tuna supplies through contractual and equity-sharing agreements with "independent" tuna fleets or by maintaining "corporate" tuna fleets. However, during the late 1970s and early 1980s, the size of the international tuna purse-seine fleet and the number of nations involved in tuna purse seining increased dramatically. As the major market for the harvest of this growing international tuna fleet, the U.S. tuna companies were in an increasingly strong buyers' position. With so many new fishermen, the threat of shortages of raw tuna greatly diminished. As a result, these companies divested themselves of their corporate U.S. fleets, entered into fewer long-term contractual arrangements with independent U.S. fleets, and began procuring raw tuna supplies on the international market from the lowest bidder. During this period, many U.S. tuna vessels were sold to nations with lower fuel and labor costs and more advantageous tax climates. This conduct put independent U.S. tuna boats at a competitive disadvantage in terms of harvesting costs during this period when the growing global tuna harvest was holding down raw tuna prices and canners had many new supply sources. According to IATTC and NMFS data, the U.S.-flag high-seas tuna purse-seine fleet operating in the ETP declined from 124 vessels in 1971 to 36 vessels in 1985. In 1991 the fleet consisted of only 11 vessels. Eight had moved to the western Pacific fishery and 6 of the 11 remaining in the ETP had licenses to fish in the western Pacific (NMFS, unpublished data).

Tuna-Processing Operations

Until the 1980s, significant differences existed in the canned tuna products from different nations. However, during the early 1980s, high-priced canned tuna in the United States and abundant low-cost raw tuna supplies from the western Pacific and Indian Ocean fishing grounds attracted many Asian nations to begin processing tuna to meet U.S. canned tuna standards. Today, most nations that process canned tuna for export meet standards for the U.S. market. This increase in the homogeneity of global canned tuna production during the 1980s and the increase in cannery production around the world greatly expanded the sources of canned tuna products for the U.S. market. U.S. companies could choose to import raw/frozen or canned tuna to meet their market requirements and began reevaluating their investments in tuna processing as well as harvesting operations. The change also placed U.S. tuna canneries in direct competition with foreign canneries who had many cost advantages and produced canned tuna that was acceptable to U.S. retailers and consumers. From 1975 to 1989, canned tuna imports increased from 8.9% (NMFS, 1980b) to 35% (U.S. International Trade Commission, 1990) of the U.S. market, and all of the major U.S. tuna companies were putting their nationally advertised labels on imported canned products.

The same general process is used to clean, cook, and can tuna in most parts of the world. The economics of tuna processing differs substantially from nation to nation, however, because of differences in direct wage rates, the costs of worker health and safety requirements, environmental regulations, and tax and trade concessions. During the 1960s and 1970s, U.S. tuna companies relocated most of the canning operations from the U.S. west coast to offshore U.S. sites in American Samoa and Puerto Rico to take advantage of favorable economic conditions including relatively inexpensive labor and tax advantages offered by commonwealth and territorial governments. During the 1980s, these companies began shifting from these offshore U.S. territories to Asian sites to take advantage of even cheaper labor and less costly worker benefits and environmental restrictions.

As competition from foreign producers of canned tuna has increased and profits from tuna canning have declined, most U.S. tuna companies have pulled out of the canning industry. As of March 1990, only one small tuna cannery is still operating within the United States; Starkist, with about 36% of the U.S. market (Iverson, 1987), is the only U.S. company still operating in American Samoa or Puerto Rico. Van Camp Seafood (Chicken of the Sea), with about 20% of the U.S. market (Iverson, 1987), was sold in 1988 to an Indonesian company, and Bumble Bee, with about 15.5% of the U.S. market (Iverson, 1987), was sold in 1989 to a Thai company. Brand-name recognition will continue to be important in the U.S. market, but the distinction between foreign and domestic canned tuna in the eyes of U.S. consumers and U.S. tuna wholesalers and retailers no longer exists. Imported canned tuna, which already accounts for 35% percent of the U.S. market (U.S. International Trade Commission, 1990), is likely to increase as a percentage of U.S. canned tuna supplies at the expense of domestic canned tuna produced in American Samoa and Puerto Rico.

Industrywide Trends

The U.S. tuna market is 31% of the global tuna market, and an estimated 70% of the U.S. tuna supply is imported in either raw/frozen or canned form (Peckham, 1989). U.S. tuna harvesters and processors have been moving away from the United States, leaving it more dependent on tuna imports. This trend will probably accelerate if any new costs are imposed on U.S. tuna harvesters or processors that are not incurred equally by foreign tuna companies. To the extent that attempts by the United States to reduce dolphin mortality increase costs or reduce productivity for U.S. vessels, these vessels will lose what little competitive advantage they have in the ETP tuna fishery and are likely to be sold to foreign investors to remain competitive. It remains unclear whether the U.S. government can (1) require U.S. tuna fishermen to operate at a competitive disadvantage; (2) subsidize fishermen to remain under U.S. jurisdiction; or (3) remove the advantage of foreign vessels by restricting access to the U.S. market. Developing engineering-based solutions to the tuna-dolphin problem should be viewed as only the first step in reducing dolphin mortality. Unless the solution is cost-effective, U.S. and foreign tuna harvesters will not be influenced easily to employ new dolphin-saving equipment or procedures or to avoid dolphin-associated fishing.

The Effects of Recent Changes

The committee notes that two recent changes are likely to affect the tuna industry and dolphin mortality. These changes are the inclusion of migratory tuna in the 1990 amendments to the Fishery Conservation and Management Act and the decision of three major tuna canneries not to sell tuna in the United States unless it is certified "dolphin-safe." Tuna from the ETP is considered dolphin-safe by the canneries if it is (1) caught by purse seiners of greater than 400 tons capacity and is caught on a fishing trip where an IATTC or NMFS observer certifies that no dolphins were intentionally encircled; (2) caught by purse seiners of less than 400 tons; or (3) caught by any method other than purse seining. Tuna caught elsewhere is considered dolphin-safe regardless of method of capture. In addition, the Dolphin Protection Consumer Act of 1990 (P.L. 101–627, Title IX) sets forth requirements for labeling of tuna as dolphin-safe and requires the Secretary of State to "immediately seek, through negotiations and discussions with appropriate foreign governments, to reduce and, as soon as possible, eliminate the practice of harvesting tuna through the use of purse seine nets intentionally deployed to encircle dolphins."

The committee believes the effects of the changes could be substantial, with the greatest impact probably on U.S. boats (see Sakagawa, 1991), but understanding them will take careful analysis. It is also probable that the effects will take some time to work their way through the industry. For these reasons, the committee has not attempted to analyze the effects of the changes, but believes such an analysis would be useful to policymakers.

3

Background on Fishing Gear

PURSE SEINES

Purse seines are used around the world and they are important in the capture of many schooling species of fish (Sainsbury, 1971). They were first used by Southern California tuna fishermen in 1916. The U.S. bait-boat fleet converted to purse-seine gear between 1957 and 1963; the design of the gear and the conversions necessary to change a vessel from bait fishing to seining are well documented by McNeely (1961).

The process of purse seining any species of fish involves the encircling of the school with a long net to form a circular wall of netting (see Figure 3-1). The net must be deep enough to discourage escape underneath it and the encircling must be done rapidly enough to prevent escape before the ends are closed. The purse seine is rectangular, typically much longer than it is deep. A seine is approximately 1 mile long and 600 feet deep (IATTC, 1989c). The top edge, or corkline, is kept at the surface by numerous floats attached along its length. The lower edge of the net, or leadline, is weighted by lead or chain to pull it down vertically. The amount of weight used and the flotation power of the net determine the sinking rate and are important in ensuring the capture of a school. The amount of weight also determines the flotation requirements, because the corkline must be kept at the surface to prevent the fish from escaping over the top of the seine.

Once the school is located, the skiff is released from the stern of the vessel, with one end of the net (known as the "ortza") attached. The skiff anchors this end of the net while the seiner encircles the targeted school and rejoins the skiff. The ortza is transferred to the vessel and made fast, thus closing the circle once the towline has been retrieved. At this point the net forms a vertical cylinder around the school of fish. To allow closure of the bottom of the seine, a series of rings are attached to the leadline through which a purse line is run. During the pursing operation, this purse line is pulled in from both ends, choking off the bottom of the seine. When the seine is completely pursed and the rings are alongside the vessel, the process of hauling in, or "drying up," the net can begin.

The remainder of the normal purse-seining operation involves "sacking up" the catch or reducing the volume of water inside the net until it is possible to bring the catch aboard using a large dip net called a "brailer." This is done through a process of bringing most of the net aboard, leaving only a small sack of reinforced netting in the water to confine the catch for brailing. Once the fish are removed, the remainder of the seine is brought aboard and made ready for the next set.

The large size of the seine and fact that the net is only in the water after a school of fish have been sighted all contribute to the efficiency of the purse-seining operation and its high productivity rates compared with other methods of fishing. The particular purse-seine gear and methods that have evolved in the Pacific tuna fishery, in comparison to other purse-seine fisheries, have epitomized the modern, efficient fishing operation.

The sacking-up process is complicated considerably during sets on tuna associated with dolphins. To minimize dolphin mortality, a variety of gear modifications and operational procedures have been adopted. The intent of these changes is to remove the dolphins from the purse seine after it has been pursed but before it is dried up for brailing. While relatively effective at its intended purpose, the process can add to the time required to complete a set and adds to the risk of losing tuna.

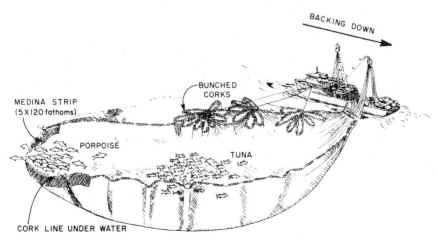

FIGURE 3-1 A purse seine showing the backdown and release of dolphins. Printed courtesy of the National Marine Fisheries Service.

The Backdown Process

The gear modifications and operational techniques mentioned in Chapter 1 are for the backdown process, in which the seine is pulled out from under the herd of dolphins. The dolphins are quite capable of jumping over the corkline and escaping, but for some reason they will not do so; therefore, the backdown was developed. In this process, the seiner is put into reverse after about one half of the net has been rolled aboard. This has the effect of forming the net into a long, narrow channel and of causing the corkline at the apex of the channel to sink. The dolphins are herded toward the apex where they can then swim away over the sunken corks.

The basic techniques of backdown were developed by the industry (the procedure was introduced by Anton Misetich in 1958). Most of the NMFS and industry effort during the 1970s was spent in attempting to perfect the design of the seine and the detailed techniques of its use for improved backdown effectiveness. The modifications that came out of that effort resulted in significant reductions in dolphin mortality (Coe et al., 1984). Modifications include the following:

• *The Medina panel.* Normal netting used in the purse seine caused frequent dolphin entanglements when they came in contact with the seine during backdown. Panels of smaller mesh netting in the portion of the net that becomes the backdown channel prevent entanglements.

• *Use of the skiff to prevent collapse of the seine.* To keep the net from collapsing while dolphins remain inside, the net skiff is used to pull the seiner to starboard, away from the seine.

• *Use of speedboats to prevent net collapse.* The speedboats, whose principal purpose was to prevent the dolphins from getting away from the seiner, are equipped with towing bridles and can be used to tow on the corkline if the net threatens to collapse.

• *Use of rafts and swimmers to effect release.* A crewman in an inflatable raft is deployed within the net to herd the dolphins toward the release area near the apex of the backdown channel to prevent them from swimming back toward the seiner and to help in the manual rescue of trapped or entangled animals.

• *Optimized set orientation and backdown maneuvering.* Guidelines to aid in determining the best orientation of the set with respect to wind conditions and proper rudder, bow thruster, and skiff controls were developed to minimize the chance of the net billowing, a situation called "canopying."

• *Pear-shaped snap rings.* These devices, which can save up to 15 minutes in beginning the backdown process, are described in Chapter 7.

The combined effect of these improvements has yielded a significant reduction in dolphin mortality since the mid-1970s (see Figure 6–1). Appendix 1 includes the gear and operational regulations that U.S. tuna purse seiners must obey when fishing for tuna associated with dolphins. Compliance with these regulations is observed and enforced. Other nations have similar regulations.

OTHER FISHING METHODS

Although purse seining for yellowfin tuna associated with dolphins in the ETP is the primary focus of this report, other methods, including additional purse-seining modes, are used for catching yellowfin. Some of these methods are known to kill dolphins, as are other techniques of fishing for other fish species (Northridge, 1984, 1991). The following are the most important.

Longline Fishing

Longline fishing is used to catch yellowfin in many parts of the world, including the ETP. The fishing lines are long and have baited hooks suspended along them, usually at some depth. The ends are buoyed. Usually, the lines are set for periods of a few hours to a day or two. Catch rates for longliners are considerably lower than for purse seiners, but the fish caught are, on average, larger than those caught in purse seines, and their economic value for some specific markets (e.g., Japan) is higher. Without a major expansion of the demand for this type of tuna (mainly used for fresh sashimi), redirecting all effort toward this mode of fishing is not economically viable, because it is considerably more expensive than purse seining.

Log Fishing

Tuna are attracted to floating objects such as logs and debris. These floating objects are referred to collectively as logs. Sometimes, artificial fish-aggregating devices (FADs) are set in the ocean to attract tuna. The tuna that collect around the logs are then caught by means of purse seines.

School Fishing

Sometimes, a school of tuna not associated with dolphins or a floating object can be detected from signs on the surface of the water. Schools moving energetically close to the surface disturb the water, which sometimes appears to be boiling or affected by a breeze. Frequently, the presence of birds is a further clue to the presence of a "boiler" or "breezer." In other cases, a school swimming close to the surface is detected as a "black spot" from the vessel or helicopter, or the tuna are seen jumping. Setting on all types of unassociated schools is known as school fishing.

4

Biology and Ecology of Yellowfin Tuna

DISTRIBUTION AND STOCK STRUCTURE

Yellowfin tuna are distributed worldwide in all warm seas except the Mediterranean (Cole, 1980). In the Pacific Ocean, their range extends from approximately 40° S to 40° N latitude. Temperature is an important determinant of the distribution of adults, particularly at the northern and southern limits of their range. Although yellowfin have been captured in waters as cold as 15° C and at least as warm as 31° C, commercial concentrations generally occur in waters between 20° and 30° C.

No distinct geographic break in distribution is evident across the Pacific. However, tagging studies indicate that it is unlikely that much intermingling of yellowfin occurs in the western, central, and eastern regions of the Pacific. Suzuki et al. (1978) postulate that there are three stocks of yellowfin tuna in the Pacific and possibly subpopulations within these stocks. Discriminant analysis applied to morphometric data from yellowfin collected from a number of locations in the ETP indicate that fish sampled north of 15° N to 20° N differ from those collected south of this region (Schaefer, 1989). Additional evidence for the existence of separate stocks within the ETP based on analysis of morphometric characters and gill raker counts is presented by IATTC (1989d). If there are indeed separate stocks, then management must focus on the stocks rather than on the species as a whole.

PHYSIOLOGY, GROWTH, AND LONGEVITY

Yellowfin possess a complex vascular network (rete mirabile) that acts as a countercurrent heat exchanger and allows them to maintain body tempera-

38

tures 1°–5° C higher than ambient water temperature. The adaptive significance of elevated body temperature and the ability to thermoregulate are incompletely known, although a number of hypotheses have been proposed (Cole, 1980).

Yellowfin are relatively short-lived and fast-growing compared with many species of fish. Age of yellowfin has been estimated by counting daily growth increments deposited on otoliths. This technique has been validated via tetracycline injection and a mark-recapture experiment for fish between 40 and 100 centimeters (cm) in length (Wild and Foreman, 1980). Although the aging method has not been validated for fish smaller than 10 cm or larger than 100 cm, circumstantial evidence suggests that the technique produces reasonable age estimates across the entire size range encountered in the ETP (Wild, 1986).

Yellowfin reach fork lengths (the distance from the snout to the fork of the tail) of approximately 40 cm at 1 year of age, 90 cm at 2 years, 120 cm at 3 years, and a maximum size of about 160 cm (Wild, 1986). Females are generally slightly larger than males between the ages of 1 and 2 years, but the pattern reverses at greater ages (Wild, 1986).

The sex ratio of yellowfin tuna is approximately 1:1 up to a length of approximately 140 cm, but thereafter the proportion of females captured in the fishery declines dramatically (IATTC, 1990a). Although the decline in females may be due to emigration from fishing areas or to a decrease in vulnerability to fishing gear, these possibilities appear unlikely because no large accumulations of females have been detected by surface or longline fleets. Wild (1986) postulated that an increase in natural mortality rate of females may account for the change in sex ratio with increasing size and further stated that the usual upper limit of female longevity in the ETP may be 3.5 years. In contrast, ages of males, as determined by Wild, ranged to just under 5 years.

ABUNDANCE AND SIZE STRUCTURE

A variety of techniques have been used to obtain estimates of absolute and relative abundance of yellowfin tuna in the ETP. An estimate of the absolute abundance of a species attempts to quantify the actual number or biomass of that species in a given area. The requirements for calculating such estimates for marine species are often impossible to fulfil, and so fishery biologists have developed and use estimates of relative abundance, which indicate the direction of changes in the size of a population rather than quantify its actual size. These estimates are simpler to calculate than absolute abundance and are correlated with it to give an idea of absolute abundance. Indices of relative abundance have been developed based on catch-per-day fished (with adjustments made for differences in fishing power among vessels) and catch-per-hour of search time (IATTC, 1989b). Cohort analysis (which uses catch-at-age data obtained from samples taken in the fishery) provides estimates of absolute abundance. Although some differences occur in estimates of abundance produced by the three methods, trends in abundance during 1968–1988 are quite similar. Abundance of yellowfin was low during 1982–1983; it increased dramatically thereafter, reaching historically high levels, and showed a recent decline (though 1988 abundance estimates still exceed the earliest average abundance) (IATTC, 1989b). During this period, annual recruitment (i.e., the number of fish of a given generation that enter the fishery in a given year) has fluctuated by approximately a factor of three. No relationship between stock size and recruitment has been detected over the observed range of population sizes. Recruitment estimated for 1983–1988 has been above the long-term average.

The size distribution of yellowfin tuna varies among modes of capture. The size distribution of fish caught as school fish shows a slightly greater mode and a broader size range than those caught in association with floating objects, whereas yellowfin caught with dolphins are significantly larger, on average, than those caught by the other two methods. The spatial distribution of fishing effort also differs by fishing mode, season, and among years. Non-dolphin sets tend to occur closer to shore than dolphin sets do, and during some periods, non-dolphin sets occur in extremely localized near-shore areas, whereas dolphin sets tend to be more broadly distributed (see Figures 1-1 and 1-2).

YIELD, POPULATION DYNAMICS, AND MANAGEMENT

The management objective for yellowfin tuna in the ETP is to obtain the maximum sustainable yield. Although catch quotas were in effect in some areas during 1966–1979, no catch restrictions have been implemented since 1979 (IATTC, 1989b). Estimated yellowfin catch in 1988 was the greatest on record, both for the IATTC's Yellowfin Regulatory Area and the eastern Pacific Ocean as a whole. Figure 4-1 shows the yellowfin and skipjack catch for 1960–1988.

The yield potential of yellowfin tuna varies greatly, depending on the size distribution of fish captured. During 1977–1981, small, medium, and large fish were all well represented in the catch, whereas during 1985–1986 the catch consisted mostly of large fish (IATTC, 1989b). The estimated yield per recruit was approximately 24% greater in 1985–1986 than in 1977–1981 (IATTC, 1989b). In general, at catch levels near or above recent levels, yield per recruit decreases as the size distribution of yellowfin captured decreases (see Figure 77, IATTC, 1989b).

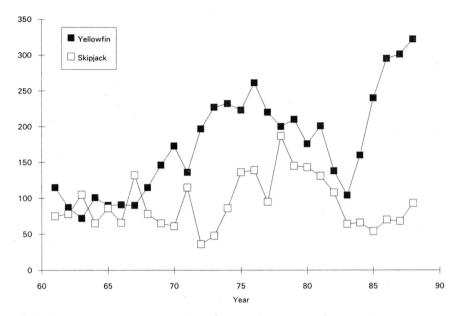

FIGURE 4-1 Eastern Pacific catches (short tons in thousands). (1 short ton = 0.907 metric ton.) Data from IATTC.

5

The Behavior of Dolphins and Tuna in the ETP

A great deal of effort has been expended to understand the impact of purse seining on dolphin and tuna stocks of the ETP. Features such as growth, reproduction, distribution, and trends in abundance of various tuna and dolphin stocks under exploitation have all been under continuing analysis for more than a decade. In addition, sporadic direct studies of dolphin and tuna behavior have been done during the course of purse-seining operations. The major impediment to doing comprehensive behavioral studies has been that such studies require work aboard seiners. While work on most other aspects of dolphin biology and mortality can go on while a seiner goes about its work, behavioral studies involve long-term dedication of a vessel to such study. Such an intensive long-term effort has not been undertaken.

The empirical experiments of fishermen (with input from NMFS) are largely responsible for the gear and methods we see today. They represent a subtle exploitation of the behavioral biology of tuna and dolphins. Such development has resulted, however, in an effective tuna capture method that continues to kill thousands of dolphins each year. For these reasons, the behavioral work of both scientists and fishermen needs to be extended.

The sole behavioral effort during actual seining operations was the Dedicated Vessel Program of 1978–1979 (see Fox and Lenarz, 1975; Stuntz, 1977; Bratten et al., 1979; Powers et al., 1979; DeBeer, 1980; Holbrook, 1980), and this program allowed only a relatively few experimental net sets to be made. Even though it was short, much of the work discussed here came from that pioneer program. In nearly every case, ideas could not be followed very far, and many tests were ended before their real promise could be assessed.

GENERAL DOLPHIN BIOLOGY AND ECOLOGY

Several species of dolphins are found in association with tuna (Table 5-1). The spotted dolphin *(Stenella attenuata)* is by far the most important from the point of view of frequency of association with tuna and use by fishermen for catching tuna. Three stocks of this species are in the ETP (Figure 5-1). The northern offshore and southern offshore stocks are considered separately; the coastal stock is subjected to very little fishing effort. The different boundaries on the figures correspond to the limits used in different assessments of abundance of the dolphin populations.

The frequent appearance of spinner dolphins *(Stenella longirostris)* in sets makes this species quite significant as well, although in almost all cases it appears in mixed herds with the spotted dolphin. Figure 5-2 shows the stock boundaries proposed for this species. Recent research by Perrin (1990) suggests that the northern and southern stocks of whitebelly spinner dolphins be merged into a single stock. In this case, every stock except the Costa Rican stock is to be considered separately. (See Perrin, 1990, for taxonomic analysis of subspecies of *S. longirostris*; he describes the Costa Rican stock as a separate subspecies.)

The common dolphin *(Delphinus delphis)* is another important species, although sets on this species are less frequent than on the previous two. There are three stocks—northern, central, and southern (Figure 5-3).

A few other species are found in association with tuna but much less frequently. These include the striped dolphin *(Stenella coeruleoalba)*, the roughtoothed dolphin *(Steno bredanensis)*, the bottlenose dolphin *(Tursiops truncatus)*, and Fraser's dolphin *(Lagenodelphis hosei)*.

TABLE 5-1 Species of Dolphin Found in Association with Tuna as Determined by Frequency of Sets on Different Species[a]

Dolphin	Percent Sets in 1988[b]
Northern Offshore Spotted	81.6
Eastern Spinner	23.7
Northern Whitebelly Spinner	8.6
Central Common	4.7
Southern Offshore Spotted	2.1
Northern Common	2.0
Southern Common	1.2
Southern Whitebelly Spinner	0.8

[a]Data from IATTC.

[b]The sum of the percentages exceeds 100 because many sets are made on mixtures of species or stocks; thus many are counted twice (e.g., almost all sets on eastern spinner dolphins also include spotted dolphins).

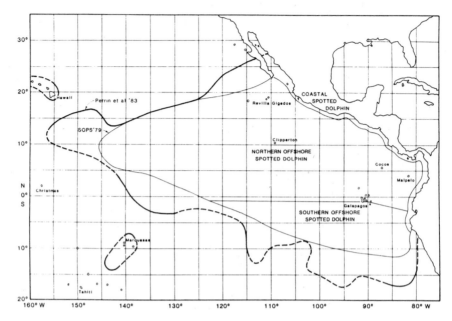

FIGURE 5-1 Known distribution of the spotted dolphin in the eastern Pacific, showing the 1979 species range and stock boundaries presented at the Status of Porpoise Stocks (SOPS) Workshop and the 1983 species range (Perrin et al., 1983). Source: Perrin et al., 1985.

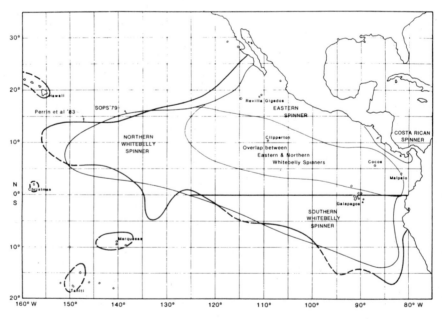

FIGURE 5-2 Known distribution of the spinner dolphin in the eastern Pacific, showing the SOPS 1979 species range and stock boundaries and the 1983 species range (Perrin et al., 1983). Source: Perrin et al., 1985.

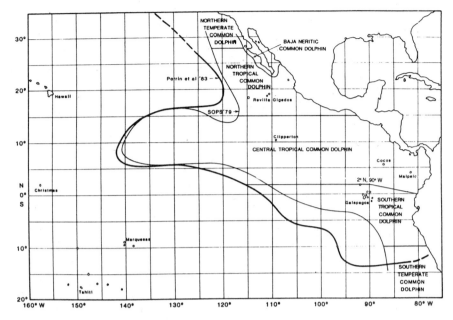

FIGURE 5-3 Known distribution of the common dolphin in the eastern Pacific, showing the SOPS 1979 species range and stock boundaries and the 1983 species range (Perrin et al., 1983). Source: Perrin et al., 1985.

THE TUNA-DOLPHIN RELATIONSHIP

The bond linking tuna and spotted dolphins is a remarkably strong one. It may persist through much or all of the seining operation. This bond means that during seining, tuna and dolphins continue to associate so tightly that to catch dolphins is also to catch tuna.

It has been a subject of debate whether the tuna-dolphin relationship is one-sided or symmetrical (i.e., is the attraction mutual or is one species attracted to the other?). The point is worthy of attention not only because the nature of the relationship may predict the course of fish accumulation but also because it will affect choices made on whether to manipulate one partner or the other. If, for example, the dolphins are central in the association and the tuna are followers, an operation that attempts to release dolphins through the net perimeter is likely to result in tuna following.

Some modestly convincing evidence indicates that the dolphins, especially the spotted dolphin, are the central species in the association and not the tuna. That is, the attraction involved is of tuna for dolphin and is neither a mutual bond nor a dolphin-to-tuna bond. This view is prevalent among fishermen and is supported by the following observations.

• No evidence suggests that dolphins protect tuna, rather the association appears passive on the part of the dolphin. It seems reasonably clear that even though fishermen repeatedly remove tuna from dolphin herds, the two species continue to reaggregate daily and thus to sustain the fishery.

• The seining operation itself strongly suggests that tuna follow dolphins. The seining operation works because it can corral the air-breathing dolphins by chasing and encircling them in a wake spiral. The dolphins must breathe frequently, especially under the high-exertion conditions of the chase, and hence they must stay near the surface where the seining operation is effective. The tuna are under no such constraint, and unless they adhere to the dolphin herd, it is unclear how seining catches them.

• During bait-fishing days (jack-pole fishing), dolphins would stay near tuna vessels. Bait fish were tossed into the water (chumming) to hold fish near the boat where they could be hooked. If the dolphins remained near the vessel, tuna could be caught, but if the dolphin herd moved away, so did the tuna (H. Medina, El Cajon, Calif., personal commun., 1990).

• During seining, if a dolphin herd becomes fragmented during the chase and part succeeds in escaping, the escaped dolphins often carry tuna with them. Conversely, if the maneuver successfully contains all of the dolphins, then the fish school tends to remain in the net (H. Medina, personal commun., 1990).

• Tuna have been observed underwater in a purse seine following dolphins (Norris et al., 1978) but dolphins do not seem to follow tuna.

One possible cause for the tuna-dolphin bond is that the echolocation system of the dolphins allows them superior environmental surveillance related either to food finding or to protection or to both. Considering the moderate visibility in daytime surface waters of the ETP, dolphins should be able to detect food schools by echolocation at least three to five times farther away than a fish could detect them by eye. Since light is extinguished rapidly with increasing depth (McFarland and Loew, 1983), such a disparity should be greatly heightened during even rather shallow dives by the combined fish and dolphin herd.

Of course, food detection may be based on a sense other than vision. Many fish have very sensitive chemoreceptors (taste and smell) and can detect some compounds at extremely low concentrations (Hasler, 1966). Dolphins have highly developed organs of taste but do not have a sense of smell, so it is unclear if one species has an advantage in locating food with these means. It is likely, however, that the dolphin is superior to the tuna in terms of detailed environmental inspection beyond the reach of sight because of its echolocation.

FEEDING BEHAVIOR

Both spotted dolphin and yellowfin tuna feed predominantly on aggregations of prey that are thought to school in or above the thermocline, although each species eats a slightly different aggregation of species from this food source. Both are thought to feed primarily during daylight or twilight (Alverson, 1963; Perrin et al., 1973; Scott, 1991), although spotted dolphins may also feed at night to some extent (Leatherwood and Ljungblad, 1979; Scott and Wussow, 1983).

The spinner dolphins that often are seined with the spotted dolphin and tuna are a diving species that feeds primarily at night on small mesopelagic prey that typically concentrate in daytime below the thermocline. Some of the dolphin and tuna food species rise to the surface or close to the surface after dark (Fitch and Brownell, 1968; Perrin et al., 1973). Because prey that are not known to reach the surface are in the spinner's diet, the spinners are believed to dive to feed (Fitch and Brownell, 1968) in tuna areas, as they do elsewhere (Norris and Dohl, 1980a). Norris et al. (in press) suggest that the open-ocean spinner dolphin of the ETP aggregate with the spotted dolphin as part of a diurnally symmetrical pair of species that may be protective of the partners, one species resting while the other feeds.

The association of large tuna, but not small ones, with spotted dolphins is probably due to physical constraints. Small tuna, whose swimming costs at a given velocity are higher than those of larger dolphins, simply may not be able to sustain such a relationship. Body length is a major determinant of the energy expenditures of swimming (Lang, 1966), just as hull length is for ships. Large tuna have body masses and lengths similar to the dolphins they associate with. In addition, there are behavioral, ecological, and probably physiological requirements for the partners in this association, not all of which are understood.

BEHAVIOR OF DOLPHINS AND TUNA IN THE TUNA-SEINING OPERATION

The events in a tuna purse-seine set involve a subtle interplay between the maneuvers of the vessel and the sensory capabilities and behavior of the dolphins and the fish. The fishermen have evolved a means by which they have turned an empirical understanding of the behavior of the dolphins and tuna into a remarkably effective maneuver to catch fish. What happens in a set is worth exploring in some detail since these events also frame the causes of most dolphin injury or mortality. These events also might be used to find ways to release or maneuver dolphins in the net without losing tuna.

The entire ETP yellowfin-tuna seine fishery depends on the unwillingness of yellowfin tuna and the dolphin species most commonly associated with them (spinner, spotted, and common dolphins) to spend much time below the shallow thermocline of this area of ocean during daylight hours (Au and Perryman, 1977; Au et al., 1979; Sharp and Dizon, 1978).

This association has been observed—and used by fishermen—in other oceans. Reports of association of yellowfin tuna with dolphins in other oceans include Cayre et al., 1988 (eastern Atlantic); Coan and Sakagawa, 1982 (eastern Atlantic); Levenez et al., 1980 (eastern Atlantic); Maigret, 1981, 1990 (eastern Atlantic); Mitchell, 1975 (eastern Atlantic); Pereira, 1985 (eastern Atlantic); Simmons, 1968 (eastern Atlantic); Stretta and Slepoukha, 1986 (eastern Atlantic); Stuntz, 1981 (central Pacific); Living Marine Resources, Inc., 1982 (Gulf of Mexico); Leatherwood and Reeves, 1989 (Indian Ocean); Montaudouin et al., 1990 (Indian Ocean); Potier and Marsac, 1984 (Indian Ocean); Northridge, 1984 (many areas); Caldwell and Caldwell, 1971 (western Atlantic); Dolar, 1990 (western Pacific); and Pacific Tuna Development Foundation, 1977 (western Pacific).

Some of these mention only observations of the association, and in one or two cases personal communications from third parties are mentioned; others include data on the sets made on dolphins.

In general terms, the frequency of the tuna-dolphin association seems to be much lower in the other ocean areas; for example, off the coast of West Africa the percentage of sets made on dolphins ranges from 0.4% in the Cape Lopez area to 4.7% off Senegal. There are no precise estimates of mortality for any other ocean area, and the proportion of trips with scientists or observers aboard is extremely low. The only mentions of mortality rates are in Levenez et al. (1980, 15 dolphins per set), Dolar (1990, 5 dolphins per set), and Mitchell (1975, several hundred dolphins in one set), but the data were not collected by independent observers; they came from surveys of fishermen and other sources.

In the ETP tuna grounds, a seine that is made deep enough can hang downward through the thermocline, and fish and dolphins will be reluctant to escape through the open bottom of the net. A possible exception is the so-called "untouchables," an apparently regional collection of dolphin herds that may have learned to escape seines before pursing takes place. Some dolphin species, such as the bottlenose dolphin, routinely dive to safety through still open seines. The thermocline of the ETP is shallowest near the Central American coast and deepens to the west, where successful seining of dolphin-associated tuna becomes more difficult at about 80-meter depths (Sund et al., 1981).

Sighting Dolphin Herds

The dolphin herds under which tuna may aggregate are usually first located by ships' lookouts using high-powered binoculars. One of the cues is the sighting of bird flocks circling near the horizon. Plunging or surface-feeding birds such as boobies and terns are often accompanied overhead by frigate birds, which, because their feathers are wettable, are restricted to stealing prey from other birds or to catching flying fish flushed from the water by the animals below. Fishermen have reported that the general size of a dolphin-tuna group below a bird flock can be predicted by the number of frigate birds flying over it (Norris and Dohl, 1980a). This general relationship between bird flocks and marine mammals appears to be widespread (Au and Pitman, 1988). The bird flocks can be detected also by sensitive radar.

The Chase

Some evidence (Au and Perryman, 1982) indicates that some dolphin herds sense and turn away from vessels as far away as seven nautical miles. The same evidence suggests that evasive maneuvers by the dolphins may quickly increase when a vessel turns toward such herds. Once a dolphin herd has been located, the seiner launches its speedboats, which race into position outside the fleeing animals and then move ahead and turn them while the seiner follows. The entire entourage changes the dolphin's swimming path into an inward bending arc. Both the speedboats and the churning vessel leave strong wakes and produce much underwater noise, which are thought to turn the dolphins (Norris et al., 1978). Shortly, the arc becomes a circle with the vessels now curving around in the second turn of the set. The average duration of the chase is one-half hour (M. Hall, personal commun., 1991).

The dolphins sometimes attempt to cut across ahead of the bows of the fast-moving seiner and to escape out through the relatively clear water ahead of the vessel. At the slightest hint that they might attempt this, the speedboats are dispatched ahead of the seiner to race in tight noisy circles in the clear water, beating it to a froth, in attempts to force the dolphins back into the wake spiral.

The Set

As the seiner spirals inward, a diameter is reached for which the circle's circumference corresponds to the length of the purse seine plus the towline. The net skiff is then released off the seiner's stern, pulling the net with it. The timing of this release is made by the skipper on the basis of existing wind, wave, and current conditions. Typically, the intent is to close the circle of net at the completion of a full turn and have the seiner complete the turn on the down-wind side of the net.

If, after the pickup of the skiff-end of the net, there is a gap at the stern of the seiner, the seiner backs against the net to close the circle; once the net is fully closed the pursing operation begins. After the net is pursed and no fish or dolphins can escape, net retrieval begins as the net is drawn up through the power block on the boom and stacked on deck.

Before the net circle is closed, dolphins seem reluctant to penetrate the wake even though it is unlikely to extend downward much more than 20 meters. A speculation is that dolphins' echolocation cannot easily penetrate the wake bubbles and that this sensory deficit may contribute to the mammals' spiral course. Free bubbles are exceptionally powerful reflectors of echolocation pulses. At resonance, the scattering and absorption cross-sections of a typical bubble at sea theoretically is of the order 10^3 times its geometrical cross-section (Clay and Medwin, 1972). Further, bubbles of the most effective scattering size for dolphin echolocation (about 60 micrometers in diameter; Glotov, 1962) may persist for long periods while smaller bubbles dissolve quickly and larger bubbles rise quickly to the surface.

Net Closed and Pursed

When the net circle is completed and pursed, the dolphins, far across the net from the now stationary seiner, appear as a slowly milling herd, located with its outer edge about 70 meters inside the far corkline and about 450 meters (or more with recent, larger nets) from the vessel (Norris et al., 1978). These netted herds move slowly, even though many individual dolphins within the herd are obviously in fairly rapid motion and chases between dolphins are frequent. Two distinct activity classes of dolphins are often present. These classes are called rafting animals and active animals. No specific species or age groups of dolphins can be associated with these two activity classes (Norris et al., 1978; Pryor and Kang, 1980). Rafting animals hang quietly and can be seen throughout netted herds. Rafts include passive columns of animals as deep as 20 meters below the surface that may from time to time rise to the surface (Pryor and Kang, 1980).

Rafters tend to be rather closely bunched together, though mostly out of actual contact with one another except for fin-tip contact between some pairs. A better understanding of rafting might contribute to understanding the problems and possibilities of release.

The herd of dolphins with many members in vigorous motion tends to remain at a single location in the net circle. The active animals tend to dive somewhere near the periphery of the herd, swim back under it, and resurface near the far edge, to turn back and dive again, back and forth, back and forth. This behavior defines what Norris (1991) has dubbed a "teacup formation," in which passive dolphins are to some extent left above and within the excursions of these more active animals, and in a fluid way the active animals roughly define the boundaries of a "cup."

Entire herds of dolphins, including rafters and more active animals, can be moved within the net circle by bringing strongly aversive stimuli near them. If, for example, a speedboat, moving by means of its noisy engine, approaches the corkline near such a herd, the entire dolphin herd has been observed to move slowly away (Norris et al., 1978). Speedboats also are able to move the herd, whereas sometimes swimmers and rubber rafts deployed inside the net cannot. The trapped dolphins tolerate swimmers moving near them, without evident flight responses. Such behavior would be unthinkable in an unrestrained herd. Encircled dolphin herds probably can be maneuvered to specific locations in the net circle to allow release efforts.

Norris et al. (1978) observed that aggression is a notable feature of active non-rafting dolphins. Aggression was also typical of patterns described in another study by Norris et al. (1985). In undisturbed spinner dolphin herds studied elsewhere in the Pacific Ocean, aggressive patterns were observed much less frequently (Norris et al., 1985).

Both Norris et al. (1985) and Pryor and Kang (1980) reported the prevalence of male coalitions in trapped dolphin herds, which has also been reported for undisturbed Hawaiian spinner herds. These coalitions consist of small, closely coordinated groups of dolphins that move as clearly distinguishable subgroups through the trapped herds. Their sex is obvious for spinner dolphins but much more cryptic for spotted dolphins (Perrin, 1972).

A protective role for these coalitions has been suggested because they frequently interposed themselves between observers and the rest of the herd, even in undisturbed herds (Norris et al., 1985). Such coalitions are active in tuna nets before backdown but have not been reported to remain during backdown. Similar male coalitions of bottlenose dolphins have been observed to cut out and sequester sexually ready female bottlenose dolphins (Connor, 1987). Although bottlenose dolphins are seldom associated with tuna or involved in the ETP tuna fishery, this observation strengthens the probability that the male coalitions have a protective role as part of their purpose.

These features indicate an organization of dolphin herds that persists early in a set and that might be utilized to assist release. If given an opportunity to escape while this protective structure is still in place, dolphins may be able to help themselves escape. In the crowding of the backdown channel, however, such social arrangements may be crowded out of existence.

6

Dolphin Mortality and Abundance

ESTIMATING INCIDENTAL DOLPHIN MORTALITY

In the early years of the purse-seine fishery, data on dolphin mortality were not collected systematically. Data are even marginally usable for only three of the several thousand fishing trips made between 1959 and 1970. In 1972, NMFS began regular sampling of the U.S. fleet, then predominant in the fishery (Smith and Lo, 1983); this program continues. In the late 1970s, however, other fishing nations began to increase their fleets, and IATTC started an international observer program, which became operational in 1979. Until recently, having an observer on board every trip undertaken by the tuna fleet was considered too costly, which left the sampling of a proportion of these trips as the only practical alternative. In all cases, observers at sea gather information on dolphin mortality, the use of equipment and procedures to reduce dolphin mortality, sightings of dolphin herds, and other biological and environmental data. Observers are debriefed at the end of their trips to clarify data and correct errors. The data are then subjected to several quality-control procedures.

Most of the following discussion is excerpted from a document prepared for the Tuna-Dolphin Workshop organized jointly by NMFS and IATTC in San Jose, Costa Rica, in March 1989 (IATTC, 1989c). Methodological considerations, factors affecting total mortality estimates, and estimates for 1959–1990 are covered.

Methodological Considerations

Sampling Design

Given the nature of the problem, the estimation method used must compute an average mortality from a sample and then extrapolate these results to the whole population from which the sample was taken. Several options are available for sampling.

• Compute the average mortality per day at sea and extrapolate to total fleet days. Days at sea and fishing activity correlate well, but the correlation is far from perfect, because days at sea are spent in activities other than fishing, such as running to the fishing grounds, waiting for storms to pass, or drifting because of a malfunction. Indeed, a vessel may spend a whole trip fishing on logs or schooling tuna (i.e., not fishing on dolphins), but the days at sea will be added for the extrapolation. The total number of fleet days is obtained from departure and arrival records for all vessels. Location of fishing is not known; therefore, spatial considerations are not possible. NMFS uses this method for its current (real-time) estimates of dolphin mortality for the U.S. fleet.

• Compute the average mortality per set on dolphins and extrapolate to total fleet sets. By using the sets as the measure of fishing activity, many of the uncertainties or ambiguities of days at sea are eliminated. By using sets on dolphins, all the fishing effort directed toward fish not associated with dolphins is eliminated. Effort can be differentiated according to the dolphin species involved and only the right share of the total effort can be applied in the extrapolation. Knowing the locations of the sets allows the use of spatial stratification.

• Compute the average mortality per ton of fish caught, either yellowfin only or the sum of yellowfin, skipjack, and bigeye *(Thunnus obesus)* tuna and extrapolate to the total catch of tuna by the fleet. This method has the same advantages as the previous method. An additional advantage is that the mortality in a set is correlated with the size of the catch. The main drawback is that the catch in each set is an estimate, creating a new source of error.

Ratio Estimates

Estimates of mortality per unit of fishing effort (days, number of sets, weight of fish caught) are known as ratio estimates. Ratio estimates are frequently biased, that is, they tend to systematically over- or underestimate. Several options have been proposed to reduce this bias. Some have suggested modifications to the basic formula used to compute the ratio (Mickey, 1959; Pascual, 1961; Tin, 1965; reviewed in Rao, 1969). Another alternative is the use of resampling techniques that may help reduce bias.

The procedure called bootstrapping (Efron, 1982) consists of sampling repeatedly with replacement from the data base of trips. The repetitions allow the calculation of standard errors without making a set of assumptions about the distribution of the variables. In addition to being useful for estimating variances and confidence intervals, the bootstrap procedure may reduce bias in the estimates, and for these reasons it was introduced into the estimation methodology.

The Observer Effect

For purposes of estimating mortality, vessels are assumed to fish in the same areas, make similar proportions of the different types of sets, and use similar gear and procedures to reduce dolphin mortality whether or not they carry an observer. The validity of this assumption cannot be verified; only anecdotal information has been used as evidence. Thus, little scientific evidence supports the existence of an observer effect. The only way to eliminate any potential observer effect is 100% coverage.

MORTALITY ESTIMATES FOR 1959–1990

1959–1972

Few data are available for the early years of the fishery (Lo and Smith, 1986). Former crew members who were concerned about high dolphin mortality provided data for the first two trips, and a government observer provided data for the third. Thus, the data do not come from any valid sampling design. Trips were not selected at random or according to any pattern. The accuracy of the data is questionable because no standard procedure was used to collect information, and interpretation of the data cannot be determined to be correct. Estimating the standard errors is especially difficult.

A substantial problem also exists with bias in ratio estimates when sampling coverages are low. Using statistical simulations (IATTC, 1989c), the positive bias produced by ratio estimates at low sampling coverage results in over-estimates of up to 4–5% at 5% coverage even after a bootstrap procedure has been applied to reduce bias. It is likely that the overestimates were even greater at the much lower coverage—about 0.1%—used earlier and when no procedures to reduce bias were applied.

In summary, the mortality estimates for the period before 1973 (peak values of up to 350,000–653,751 in a year by Perrin, 1968, 1969; Smith, 1979, 1983; Lo and Smith, 1986) have little or no statistical value, and the only conclusion that can be based on the data available is that mortality was very high. After a long hearing, the administrative law judge, Hugh Dolan, concluded that many errors had caused dolphin mortality to be seriously overestimated and dolphin abundance to be seriously underestimated by NMFS in the 1960s and 1970s (Dolan, 1980).

Nonetheless, these mortality data were used to calculate estimates of dolphin abundance for 1959–1970; these estimates were used later to conclude that some stocks were depleted or were at a given proportion of their original abundance.

1973–1978

Wahlen (1986) computed mortality estimates for the U.S. fleet for 1973–1978. During those years, after passage of the U.S. Marine Mammal Protection Act of 1972 (MMPA), the data collection system for the U.S. fleet improved considerably. With more and better standardized data, dolphin mortality was shown to have declined from over 100,000 dolphins per year to approximately 20,000. During that period, the fleets of other countries increased in size, prompting the creation of IATTC's international observer program.

1979–1990

Various estimates of dolphin mortality for the international fleet were produced between 1979 and 1983 (Allen and Goldsmith, 1981, 1982; Hammond and Tsai, 1983; Hammond, 1984; Hammond and Hall, 1985). From 1984 on, the methods used to estimate mortality were modified, and figures for 1979–1988 are presented in a series of publications (Hall and Boyer, 1986, 1987, 1988, 1989, 1990, in press, a,b). A review of all the figures for 1979–1987 is also available (IATTC, 1989c). Figure 6-1 shows the estimates for 1972–1990.

FACTORS AFFECTING TOTAL MORTALITY ESTIMATES

Fishing Effort on Dolphins

The catches resulting from school fishing, log fishing, and dolphin-associated fishing usually are quite different in species composition and distribution of sizes of fish caught. Sets on floating objects (log sets) usually produce large amounts of skipjack associated with small yellowfin (<70–80 cm). The fish caught in school sets are slightly larger than those caught in log sets, and the distribution of size is broader in school sets than in log sets.

Sets on dolphins yield the largest tuna caught in the purse-seine fishery; most are longer than 80–90 cm and frequently are over 100–110 cm, and almost all of them are yellowfin. The canneries pay higher prices for large

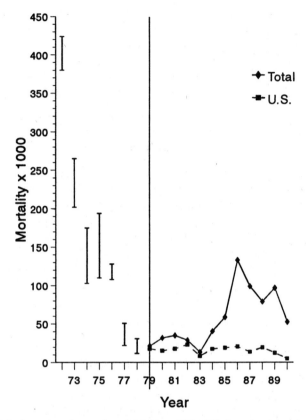

FIGURE 6-1 Estimated total mortality of dolphins, 1972–1990. Data for the period before 1979 are subject to controversy and do not permit reliable breakdown between U.S. and non-U.S. sources. See text. Data from IATTC; except total mortality data for 1972–1978; in those cases, the high range is from NMFS (1980a) and the low range is extrapolated from a figure produced by a workshop and reported by IATTC (1989c); U.S. mortality data are from NMFS (MMC, 1991).

yellowfin than for other fish because they produce greater yields with less labor. In recent years, the gap between the prices paid for large and small yellowfin has fluctuated considerably, influencing the selection of mode of fishing by the fishermen. Figure 6-2 shows the proportion of sets made on dolphins, logs, and school fish, and the proportion of the total tonnage caught in each type of set during 1979–1990. When the price gap is wide, effort concentrates on dolphins, as in 1985–1987, resulting in large increases in the number of sets and in the tonnage caught on dolphins. In previous years, when the price difference was less, the proportion of all sets made on dolphins was also much smaller. In 1988, effort on dolphins declined markedly, but in 1989, the number of sets on dolphins increased by 19.6% over 1988 to a record high (IATTC, 1991a).

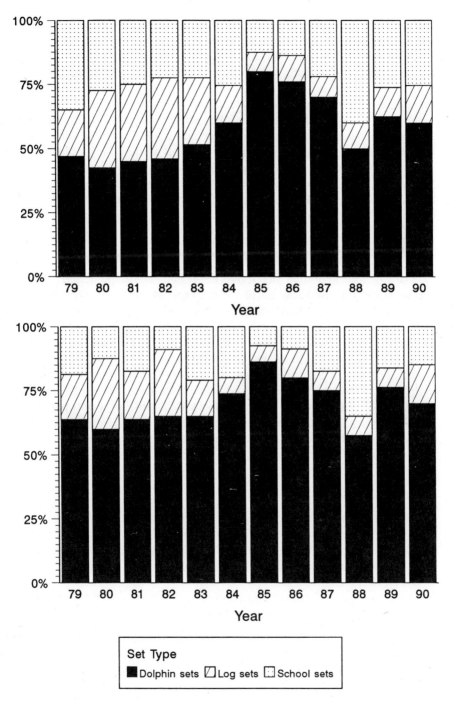

FIGURE 6-2 Proportion of sets (upper panel) and percentages of tons of yellowfin caught (lower panel) by set type. Source: Adapted from IATTC, 1991a.

The level of effort on dolphins may also be affected by the dolphin-tuna association, which may be less frequent or more temporary during certain years. A good example of this effect is the 1982–1983 period, when a very strong El Niño event disrupted the fishery. One of the main fishing areas for dolphin-associated tuna (along latitude 10 °N) was quite unproductive, and fishing on dolphins declined considerably. The changes in oceanographic conditions, especially the patterns of the temperature fields and thermocline depth, may have driven the tuna deeper and made them less available to surface gear, or perhaps the changes in currents shifted the location of the productive fishing grounds to areas not normally fished by the fleet.

Finally, due to a series of environmental, biological, and fishery factors, the abundance of large yellowfin increased to very high levels in the mid to late 1980s (IATTC, 1989a).

Factors Directly Affecting Dolphin Mortality

Many factors other than total fishing effort influence dolphin mortality, and we probably do not know all of them. Some of the most important are described briefly below. Many of these factors are interrelated. For instance, in general terms, the larger the catch of tuna in a set, the longer the set will last, the larger the number of dolphins that will be caught, and the more likely it is that the set will finish after dark (dolphin mortality increases markedly after dark) or that a malfunction will occur. Also, different species or stocks of dolphins have different herd sizes and behavior patterns and inhabit different areas. Some areas and some species or stocks are fished during only part of the year, so spatial, temporal, and species effects cannot be distinguished. This makes it difficult to identify precisely the cause-effect mechanisms that result in the higher or lower mortalities.

Catch of Tuna in Set

The size of tuna catches affects mortality per set (MPS) and mortality per ton (MPT). Clearly, years with high abundance of tuna will result in larger average catches, and therefore higher mortalities per set. In recent years, the average catch per dolphin set has been roughly twice the average in the early 1980s (IATTC, 1989b).

Vessel Captain

Frequency distributions of 1989 observed average dolphin kill per set and total annual dolphin kill by vessel captain are given in Figures 6-3 and 6-4, respectively (approximately 100% of U.S. trips and 35% of non-U.S. trips were observed in 1989). Figure 6-5 is taken from Figure 6-4 and shows the

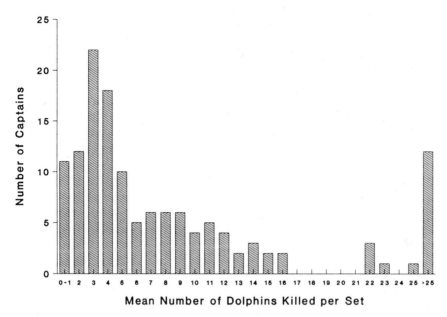

FIGURE 6-3 Average number of captains with average kill per set for 1989. Most captains killed five or fewer dolphins per set. A few killed more than 20 per set. Data from IATTC.

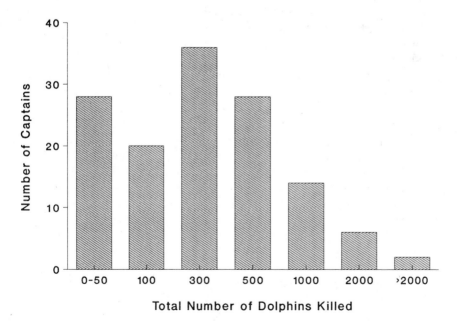

FIGURE 6-4 Total number of dolphins killed by captain for 1989. Data from IATTC.

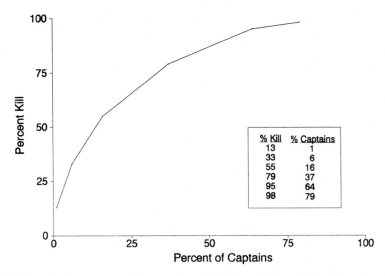

FIGURE 6-5 Percentage of captains and observed dolphins killed for 1989. Data extrapolated from Figure 6-4.

cumulative percentage of observed kill in 1989 against the percentage of observed captains that accounted for that kill. Approximately 15% of the total number of vessel captains accounted for 50% of the total dolphin kill, and conversely, 50% of the vessel captains accounted for less than 15% of the total dolphin kill. Although kill per captain is a function of the number of sets on dolphins as well as the captain's skill, skill is important. Data need to be collected and reported in such a way that these factors can be discriminated.

Species or Stock Caught

Several species or stocks of dolphins associate with tuna in the eastern Pacific. Some of the stocks occupy a large portion of the range of the fishery; others are present only in limited areas. Two sources of heterogeneity in the mortality rate of dolphins should be considered: (1) different species or stocks have different MPS values; and (2) the same species or stock may have different MPS values in different regions of the fishery.

The current mortality rates for common dolphins in sets is much higher than those for any of the other species. The average herd size for common dolphins (approximately 500) is greater than for the other stocks or species (spotted dolphins, 400; eastern spinner dolphins, 160; and whitebelly spinner dolphins, 180), and their more active behavior in the net makes them more likely to become trapped or tangled (IATTC, 1989c).

Area of Capture and Habituation

Cumulative effort and MPS values in an area are negatively correlated. Areas with more sets historically have lower MPS values; the edges of the fishery and some areas to the south that receive less effort have greater MPS values (IATTC, 1989c). This topic was discussed by Hall et al. (in press), who showed that the behavior of the dolphins seems to be affected by the fishing activity in the area in a way that leads to a reduction in mortality and that perhaps learning on the part of the dolphins was lowering the values. A spatial stratification scheme based on one-degree areas is used to increase the precision of the estimates.

Flag of Vessel

Early estimates of mortality (Allen and Goldsmith, 1981, 1982; Hammond and Tsai, 1983; Hammond, 1984; Hammond and Hall, 1985) included a stratification based on the flag of the vessel. The assumption was that countries with different degrees of regulation or enforcement were likely to have differences in mortalities. As sample sizes were insufficient to stratify by all flags, a category called "non-U.S." was used, although this group was in itself heterogeneous. However, a series of statistical tests (Hall and Boyer, 1986) for 1979-1983 showed that when fishing area was taken into account, it was not possible to reject the null hypothesis of equal MPS values for the two groups. Only two of seven areas used had significant differences, and they were of opposite sign. The conclusions of the study were that stratifying by area made more sense than stratifying by flag when samples were limited and that any differences between flags should be dealt with by using a proportional sampling design and taking the same proportion of all flags.

In recent years, however, the sampling coverage of the U.S. fleet, which is determined by NMFS, has been quite variable because of changes in policy. The rates of sampling have gone from 40% to 100%, down to 50%, and back up to 100% (IATTC, 1989c). These changes in coverage generate a bias by altering the proportion of samples from the U.S. fleet and also, in some cases, by affecting the behavior of the fleet, because the changes were made in response to reaching quotas or to lawsuits. The sampling rates for the non-U.S. fleets have been increasing, which has allowed an additional stratification by flag since 1986, and the results of these samples are added to compute the total for the international fleet. In recent years, the kill rate of the U.S. fleet has been lower than the average of the international fleet.

Time of Capture (Day vs. Night Sets)

Night sets (sundown sets) are defined as those in which the backdown procedure finishes in darkness. Because of the additional difficulties that darkness generates for the dolphin rescue effort and also for the dolphins' perception of the situation, sundown sets have much higher mortality rates than day sets. The reduction in light at dusk below the sea surface is much greater than above it (McFarland and Munz, 1975) and many animals use visual sensitivity then (McFarland and Loew, 1983). The number of dolphins left in the net after the backdown maneuver is much higher in sundown sets. According to IATTC (1989c), sundown sets, which were only 9.3% of nearly 16,000 observed sets, accounted for nearly 26% of the dolphin mortality.

Duration of Set

Longer sets have higher mortality values than shorter ones. Typically, the backdown procedure lasts about 15 minutes; however, sets with backdown lasting 40 minutes or more have mortality rates 10 times higher than sets with backdown lasting 20 minutes or less (IATTC, 1989c). Long backdowns are infrequent and may be associated either with the capture of large herds of dolphins or with a malfunction, so one cannot assume that the duration of the set is the sole cause of high mortality. However, it is obvious that the longer the dolphins stay inside the net, the more likely they are to become entangled or die trapped by a net collapse or canopy.

Presence of Strong Currents

Strong subsurface currents affect the behavior and handling of the net and may cause net collapses, one of the major causes of mortality. Sets encountering these currents result in mortality rates twice as high as sets made in normal conditions (IATTC, 1989c).

Occurrence of Malfunctions

Malfunctions of many kinds can occur during a set. Some affect the skiff, some the speedboats, some the main vessel, and some the net. Sets with minor or major malfunctions have mortality rates about twice as high as those without malfunctions (IATTC, 1989c).

Alignment of the Fine-Mesh Panel

The fine-mesh panel forms the end of the backdown channel and is effective in reducing mortality if it is properly aligned. If it is not, its efficiency decreases sharply, and MPS and MPT more than doubles (IATTC, 1989c). Proper alignment can be achieved by performing a trial set.

Use of Dolphin-Saving Procedures

The correct use of all dolphin-saving procedures usually reduces mortality considerably. Preventing net collapses and canopies, or counteracting them when they happen, also helps. Sets with net collapses or canopies have mortality rates 2 to 3 times higher than sets without such collapses or canopies. The presence of a crewman in a raft in the backdown channel is associated with a reduction in the MPS from 11.6 to 5.1 and the MPT from 0.86 to 0.36 (IATTC, 1989c).

Relationship Between Number of Vessels and Total Mortality

Table 6-1 shows the relationship between the total carrying capacities of vessels in the ETP, the carrying capacities of the U.S. fleet fishing in the ETP,

TABLE 6-1 Carrying Capacity, in Metric Tons, of the International and U.S. Tuna Purse-Seine Fleets, and Total Dolphin Mortality, 1974–1990[a]

Year	Carrying Capacity				Total Dolphin Mortality	Estimated U.S. Dolphin Mortality
	Total		U.S.			
	Seiners >400 Tons	All Seiners	Seiners >400 Tons	All Seiners		
1974	138,901	161,983	107,854	118,475	103,000–175,000	
1975	158,567	180,470	122,716	132,848	110,000–194,000	
1976	173,734	194,469	124,255	133,361	108,000–128,000	
1977	177,421	197,052	122,405	130,221	22,000–51,000	
1978	176,054	199,220	113,050	122,559	12,000–31,000	
1979	178,765	202,407	113,218	122,941	21,426	17,938
1980	181,991	203,481	117,147	124,614	31,970	15,305
1981	186,219	203,794	113,159	120,216	35,089	18,780
1982	170,786	184,893	109,100	114,840	29,104	23,267
1983	139,594	153,384	81,017	86,922	13,493	8,513
1984	114,871	125,036	48,777	52,700	40,712	17,732
1985	130,146	140,254	50,872	54,494	58,847	19,205
1986	125,740	135,065	44,664	47,645	133,174	20,692
1987	149,909	158,429	43,965	46,245	99,187	13,992
1988	154,399	163,258	46,531	49,026	78,927	19,712
1989	138,349	147,272	33,881	36,236	96,979	12,643
1990	140,910	149,279	33,220	35,412	52,531	5,083[b]

[a]Data from IATTC; except total mortality data for 1974-1978; in those cases, the low range is extrapolated from a figure produced by a workshop and reported by IATTC (1989c) and the high range is from NMFS (1980a). U.S. mortality data are from NMFS (MMC, 1991). Data for the period before 1979 are subject to controversy and do not permit reliable breakdown between U.S. and non-U.S. sources. See text.

[b]Because there was 100% observer coverage in 1990, the mortality for this year is a count, not an estimate.

and the total dolphin mortality from 1974 to 1990. The total mortality dropped significantly between 1976 and 1977, largely because of industry gear development initiatives, which were then reflected in U.S. government regulations under the MMPA.

ESTIMATES OF DOLPHIN ABUNDANCE

Estimates of dolphin abundance in the ETP have been made by NMFS and IATTC, based on sightings made from either research vessels or fishing boats. Other methods of estimating abundance, such as mark-recapture experiments, or other sources of data, for instance sightings from aerial surveys, have proved inadequate for this purpose.

The sources of data for these estimate areas are as follows:

• RVD (Research Vessel Data). Sightings from research cruises, designed and carried out by NMFS with the express purpose of estimating dolphin abundance. Holt and Powers (1982) produced estimates for 1979 using RVD alone and a combination of RVD and tuna vessel data. Since 1986, NMFS has made a series of five annual cruises to detect trends in abundance. The results are reported in Sexton et al. (in press) and Gerrodette and Wade (in press, a and b).

The RVD cruises are designed to provide a random distribution of effort and, therefore, unbiased estimation. However, given the size of the area over which the dolphin populations are distributed and the amount of searching effort (16,870 nautical miles in 1990, with 362 sightings), the estimates of abundance are imprecise. To reduce this imprecision, the data have been pooled in various ways, and as a result of this and other procedures applied to the data, RVD also produce estimates of relative abundance designed to detect trends over a period of time (Sexton et al., in press). A recent substantial revision of the methodology (Gerrodette and Wade, in press, a) reduces most of the biases; these estimates are recognized as being closest to estimates of absolute abundance (IWC, in press), and this method will be used in the future for RVD.

• TVOD (Tuna Vessel Only Data). Sightings made from tuna vessels recorded by IATTC and NMFS observers placed on board to monitor the incidental mortality of dolphins in fishing operations. The analysis of these data have been carried out mostly by IATTC and reported in Hammond and Laake (1983) for the period 1975-1982, Buckland and Anganuzzi (1988), Anganuzzi and Buckland (1989), and Anganuzzi et al. (in press, a,b). Polacheck (1987) carried out an analysis (not based on line transect methods) of the U.S. TVOD.

The sample size for the TVOD data is very large (771,534 nautical miles searched in 1990, with 20,747 sightings), but because of the nature of the fishery, searching is concentrated on areas of high density of dolphins. Unless this is taken into account in the analysis, the abundance will be overestimated. Various spatial stratification procedures have been used to alleviate these problems; see Buckland and Anganuzzi (1988) and Anganuzzi and Buckland (1989) for the most recent applications. These procedures aim at increasing the robustness of the estimates, rather than eliminating all sources of bias.

Analysis of the vessel sightings data has been based mostly on line-transect methodology (Burnham et al., 1980), which essentially estimates a correction factor to account for unseen herds of dolphins. A number of assumptions must be met in order for this method to produce unbiased estimates of abundance. These assumptions are violated to different extents by RVD and TVOD, and therefore, the statistical procedures applied in each case differ considerably. The resulting estimates are indices of relative abundance, indicating trends in abundance over the period of time in which the data were collected, rather than estimates of absolute abundance, i.e., the total size of the population. The section "Abundance and Size Structure" of yellowfin tuna contains further information on the difference between these two types of estimates.

The significance of a trend in relative abundance estimates, whether from RVD or TVOD, is determined on the basis of a linear test. A straight line is fitted to a subset of the time series of estimates, and if the slope of the line is significantly different from zero, the trend is considered significant. This test has been applied over a moving period of 5 years (Buckland and Anganuzzi, 1988; Anganuzzi and Buckland, 1989) and 10 years (Edwards and Glick, in press). However, given the imprecision of the estimates and the fact that the test does not take into account nonlinear changes in abundance, only very large changes in population size can be detected (Edwards and Glick, in press). To overcome these problems, Buckland et al. (in press) use a different method for detecting trends, based on a smoothing of the time series and a direct comparison of smoothed estimates. Additional studies are under way on the power of these tests.

Smith (1983) uses a different approach, which attempts to determine historical population levels. Estimates of abundance of three populations of dolphins (offshore spotted, eastern spinner, and whitebelly spinner) were calculated for 1959–1979, based on back-calculating estimated mortality and recruitment from the 1979 population estimates. However, as mentioned above, both mortality estimates for the period before 1972 and population estimates based on them have little or no statistical value.

The results of the analyses of RVD and TVOD are as follows:

• Offshore spotted dolphin. This is the species most frequently set on: in the 1984–1988 period, 83.9% of dolphin sets were directed at spotted dolphins, either in pure herds or mixed with spinner dolphins. Abundance estimates for the northern stock for 1986–1990, based on RVD, average 1,514,800 animals, with coefficients of variation (CV) between 29% and 36%; for the southern stock the corresponding figure is 267,400 (CV: 48–86%) (Gerrodette and Wade, in press, b). Estimated trends suggest a significant decline of the northern stock in the late 1970s, followed by a period of relative stability (and perhaps some increase) in the 1980s (Anganuzzi et al., in press, b; Buckland et al., in press). For the southern stock, there is an indication of a decline during the early 1980s. Recent high estimates for this stock suggest that the two stocks may not be as discrete as was thought, so these results should be treated with caution.

• Spinner dolphin. RVD estimates for 1986–1990 average 588,500 animals (CV: 37–42%) for the eastern spinner, and 993,700 (CV: 38–64%) for the whitebelly spinner (Gerrodette and Wade, in press, b). No significant trends were reported in 1975–1990 for eastern spinners. Trends for the whitebelly spinners also indicate a stable situation in recent years, having a possible decline in the late 1970s. In 1986–1990, 21.9% of dolphin sets involved eastern spinners and 16.5% involved whitebelly spinners; in the great majority of these sets the spinners were in mixed herds with spotted dolphins.

• Common dolphin. RVD estimates for common dolphins in 1986–1990 average 467,400 for the northern stock, 594,300 for the central stock, and 2,117,500 for the southern stock (Gerrodette and Wade, in press, b). Estimates for these populations are relatively imprecise, CVs ranging from 44% to 84%. Anganuzzi et al. (in press, b) and Buckland et al. (in press) reported a significant decline in the northern stock during the early 1980s, having an index of abundance for 1989 significantly lower than those for 1979–1981. The central stock showed evidence of a decline between 1978 and 1983 and subsequent stability. Data are sparse for the southern stock, but abundance in 1989 was significantly lower than in 1976–1978. In 1984–1988, 4.6% of purse-seine sets involved common dolphins.

• Striped dolphin. Estimates in Gerrodette and Wade (in press, b) average 172,400 for the northern stock (CV: 37–62%) and 1,313,500 for the southern stock (CV: 27–30%). Sightings of striped dolphins in the TVOD data base are too rare for a reliable estimate.

The best estimate available of the average total population of these four species of dolphins in the ETP in 1986–1990, calculated from RVD by Gerrodette and Wade, "possibly [has] a slight negative bias" and is slightly over 8,000,000 (Gerrodette and Wade, in press, b). Estimates of absolute abundance are very important because they convey the impact of the mortality within a stock due to fishing as a proportion of the total population, not as a number without any frame of reference (see Table 6-2).

In summary, both the NMFS and the IATTC studies demonstrate that none of the indicators of stock size shows any statistically significant trend in the last 5 years. The strong El Niño event of 1982–1983 affected many of the sources

TABLE 6-2 Average Population Abundance in 1986–1990, Based on Research Vessel Surveys, Incidental Mortality in 1990, and Percentage of the Total Population Killed in the Fishery in 1990

Species and Stock	Population Abundance[a]	Incidental Mortality[b]	Percent Killed
Offshore Spotted Dolphins			
Northern Stock	1,514,800	32,300	2.1
Southern Stock	267,400	1,600	0.6
Spinner Dolphins			
Eastern Stock	588,500	5,400	0.9
Whitebelly Stock	993,700	7,000	0.7
Common Dolphins			
Northern Stock	467,400	700	0.1
Central Stock	594,300	4,100	0.7
Southern Stock	2,117,500	300	0.0

[a]Data from Gerrodette and Wade, in press, b.

[b]Data from Hall and Boyer, in press, b.

of data, so it seems reasonable to use those years as a demarcation. Before 1982, and especially in the late 1970s, several stocks were experiencing large declines; since 1983 all indicators of stock size have been stable, and some appear to have been increasing (see Figures 6-6, 6-7, and 6-8). However, it should be borne in mind that the tests for trends are able to detect only fairly large changes (whether positive or negative) in the short term.

SIGNIFICANCE OF THE ESTIMATES OF MORTALITY AND ABUNDANCE

For a complete assessment of the significance of estimated mortality and abundance to dolphin populations, a better knowledge of recruitment rates and dolphin migration is needed, as well as better stock identification of individuals. At present, the eastern spinner dolphin *(Stenella longirostris orientalis)* is the subject of current concern and currently being proposed for depleted status under the MMPA. Estimates of its numbers as a fraction of prefishery numbers vary widely, because baseline estimates of prefishery numbers are low and because estimates of mortality rate—themselves very imprecise for the years before the mid-1970s—are used in calculating population numbers. Furthermore, population estimates are complicated because hybridization with surrounding populations appears to be common. Some researchers believe that hybridization may be increasing because of in-migration from surrounding populations to areas of greatest mortality.

Northern offshore spotted

Relative abundance
(in thousands)

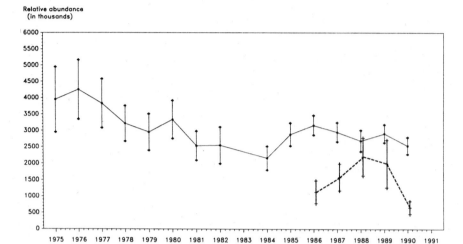

FIGURE 6-6 Relative abundance estimates for the northern offshore spotted dolphin. Solid line indicates estimates from TVOD (Anganuzzi and Buckland, 1989; Anganuzzi et al., in press, a; Anganuzzi et al., in press, b); dashed line indicates estimates from RVD (Gerrodette and Wade, in press, b). Vertical lines indicate estimate plus or minus one standard error.

Eastern spinner

Relative abundance
(in thousands)

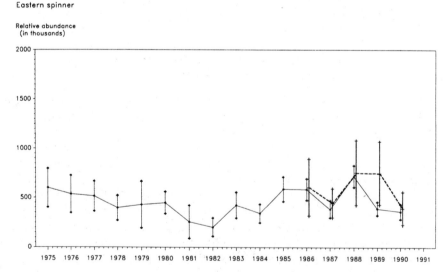

FIGURE 6-7 Relative abundance estimates for the eastern spinner dolphin. Solid line indicates estimates from TVOD (Anganuzzi and Buckland, 1989; Anganuzzi et al., in press, a; Anganuzzi et al., in press, b); dashed line indicates estimates from RVD (Gerrodette and Wade, in press, b). Vertical lines indicate estimate plus or minus one standard error.

Whitebelly spinner

Relative abundance
(in thousands)

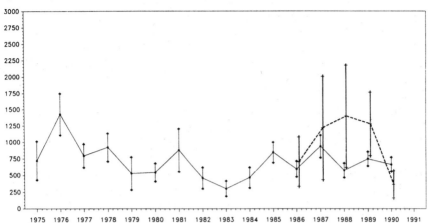

FIGURE 6-8 Relative abundance estimates for the whitebelly spinner dolphin. Solid line indicates estimates from TVOD (Anganuzzi and Buckland, 1989; Anganuzzi et al., in press, a; Anganuzzi et al., in press, b); dashed line indicates estimates from RVD (Gerrodette and Wade, in press, b). Vertical lines indicate estimate plus or minus one standard error.

Spotted dolphins, although they are the most frequent object of the seiners' efforts and suffer the greatest total mortality, seem not to have been depleted as much as the spinners, probably because their populations were much larger at the start of the fishery. The effect of the fishery seems thus far to have been of only modest importance for populations of other dolphin species of the region, although more information is needed, especially with respect to the common dolphin.

OVERVIEW: THE PAST, PRESENT, AND FUTURE

Historical Summary

Clearly, kill rates during the early years of the fishery were significantly reducing the populations of the most commonly caught dolphins. By the early 1970s, when the strictures of the MMPA were being felt, a decline in overall kill rates began under the guidance of NMFS, with the cooperation of the industry's fishery-research arm, the Porpoise Rescue Foundation. The reduction was achieved largely by training skippers to perform the backdown procedure; by helping quiescent dolphins from the nets; and by gear modifications, especially those which reduced tangling of dolphins in the net mesh.

By 1977, the mortality had declined substantially and remained low until the mid-1980s. As the overall kill rate dropped toward 25,000 per year (see Table 6-1 and Figure 6-1), other smaller factors began to be more important in producing year-to-year variations in kill than they were earlier. The El Niño years of 1982–1983 were especially important. As the fleet turned away from dolphin fishing, mortality rates dropped for a year or two, but rose again as oceanographic conditions returned to "normal" and the fishermen again fished on dolphins.

Then, as the majority of the U.S. fleet moved out of the ETP, responsibility for the attempts to reduce dolphin mortality moved primarily to IATTC. During this transition, kill rose at first as previously uncontrolled foreign vessels began to dominate the ETP fishery. IATTC developed a program to monitor and reduce the dolphin kill that began to take effect after the mid-1980s, and annual mortality again began to decline, to perhaps 25,000 in 1991. Education, the banning of sundown sets, and performance standards for skippers have been especially important in the more recent decline, and the mortality caused by the international fleet seems to be approaching those low levels again.

Maintenance and Recovery of Dolphin Populations

It is important to emphasize that the dynamics of dolphin populations under fishery pressure are different from those of the tuna caught with them. Yellowfin tuna are short-lived—5 years is old for a tuna—and each female produces millions of eggs each year. The dolphins involved live to be as much as 35 years old and produce only 12–15 young in a lifetime. Largely because of these differences, tuna populations can sustain a harvest of 20–40% of the reproductive population in a year (Francis, 1986), while the maximum safe harvest for dolphins is probably much less than 10%.

One way to approach the question of how much mortality dolphin populations can sustain and remain stable or increase is to express harvest as a proportion of net recruitment, i.e., as a proportion of the number of animals added to the population each year minus those that died. Recent estimates of dolphin abundances in the ETP vary; the best—although still imprecise— seems to be that of Gerrodette and Wade (in press, b) and is about 8,000,000 animals, with eastern spinners estimated at 600,000. Recruitment is also inadequately known (Reilly and Barlow, 1986; Smith, 1983), but an estimate of 2–6% per year seems reasonable (Smith, 1979, 1983). The lower estimate of 2% would represent 160,000 animals per year (12,000 eastern spinners per year). A kill rate of 40,000 animals per year would thus represent a kill rate of 25% or less of recruitment, almost certainly low enough to permit current dolphin populations to be stable and perhaps to increase. An annual kill of 20,000 (12.5% or less of recruitment) would probably result in substantial increases in dolphin populations. Similar calculations can be applied to individual stocks and species. Thus, with a small reduction in the 1991 mortality, it is reasonably certain that dolphin populations would increase.

It is important to note that very little is known of natural mortality in dolphins or their ability to compensate for fishing mortality as many fish populations are known to do. For example, infestation of nematode parasites in dolphins has been projected to increase natural mortality (Dailey and Perrin, 1973).

Remarkably, there is no evidence of major social disruption of dolphin populations as a result of seining and some evidence indicates that they have adjusted to it (e.g., experienced dolphins are less excitable in a tuna seine and less likely to become entangled than naive animals). Also, evidence from Hawaiian spinner dolphins shows that the membership of aggregations there can be quite variable from day to day, unlike the tight family units of many land-mammal societies (those dolphins live in "fission-fusion" societies; Norris, 1991). This means that precise membership of a herd seems not to be essential to its daily (i.e., nonreproductive) operation and that disruption by seining is less harmful than it would be if a tight social unit were being affected.

The effect of seining on dolphin reproduction is not clear. Dolphins give birth to large young that can swim rapidly soon after birth. But the young need assistance, which is provided by other members of the herd while the mother recovers from the birth. These helpers seem to fill a long-term societal role; they are not randomly chosen. Thus, disruption of a herd by seining at the time of reproduction might have adverse consequences (Norris et al., in press). Dolphins reproduce all year in the ETP but with a major peak in late spring and early summer and usually a secondary peak in early fall (Perrin et al., 1984).

The committee did not analyze different fishery-management options to conserve dolphin populations, but notes that if better information is obtained on the timing and distribution of dolphin reproduction in the ETP, time and area closures might be worth consideration. In addition, the committee notes that a complete ban on dolphin fishing or the purchase of tuna caught on dolphins is not required to ensure the survival and even the increase of dolphin populations. The committee did not attempt to judge the desirablity or validity of various management policy goals, but it is clear that any policy designed to reduce dolphin mortality or prevent it absolutely will be effective only if it is based on sound information and if most or all nations that fish for dolphin-associated tuna anywhere in the world participate in its implementation.

7

Techniques for Reducing Dolphin Mortality

The effectiveness of fishing gear is a function of its design and the way it interacts with the animals. Therefore, any successful effort to develop gear that is effective at catching tuna without killing dolphins must involve knowledge of tuna and dolphin behavior as well as good engineering and design.

SMALL MODIFICATIONS OF CURRENT METHODS

Several small modifications to the current methods of purse seining for tuna have immediate potential for reducing dolphin mortality. Each of these changes could have an incremental effect. The cumulative effect of these and other innovations could significantly reduce the impact of purse seines on dolphins.

Medina Double Corkline

In spite of dolphins' characteristic trait of jumping well clear of the water, they do not leap over the corkline to freedom. When they make an effort to escape over the corkline, typically during the backdown process, they may push against the corkline, deflect it downward, and swim out.

During backdown, the corkline surrounding the backdown channel is tight and resists being submerged. To sink the corkline under this condition, a large number of floats must be submerged. Only at the apex of the backdown channel and only when there is sufficient flow through the channel, does the corkline begin to submerge.

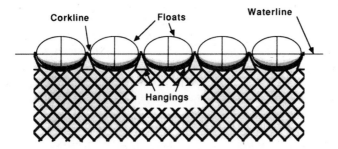

FIGURE 7-1 The conventional purse-seine corkline. Drawing by C. Goudey, Massachusetts Institute of Technology, Cambridge.

The use of a double corkline has been suggested as a way to ease the dolphins' escape (H. Medina, personal commun., 1990). Unlike a conventional corkline, shown in Figure 7-1, which runs through the center of each float, the double corkline has each float tied on individually. As seen in Figure 7-2, this rigging allows each float to move independently, accommodating dolphins' attempts to push their way over the corkline.

Rigging a seine in this way seems to present no hardship to the net builder. Hanging the netting to the corkline, in fact, would be much easier and could be done more precisely. The method has the additional advantage of allowing easy replacement of damaged floats.

Jet Boat

In the present backdown process, use is often made of an inflatable raft with an occupant trying to direct the dolphins toward the end of the backdown channel. In the past, no motorized craft could be used for this task because of

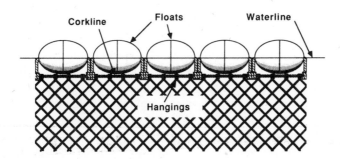

FIGURE 7-2 The double corkline. Drawing by C. Goudey, Massachusetts Institute of Technology, Cambridge.

the threat posed by the propeller. The introduction of the jet boat as a recreational watercraft now allows an alternative to the unpowered raft in controlling dolphin movements within the backdown channel and aiding in dolphin rescue. A Japanese manufacturer has introduced a fiberglass craft with a 52-horsepower engine that is propelled by a jet pump. This craft is 9 feet long and 5 feet wide, large enough for three occupants, and has a hull that sits low in the water. It can turn within its own length. A demonstration of the craft in San Diego in April 1989, sponsored by the Porpoise Rescue Foundation, indicated that it might be useful both inside and outside the net and before, during, and after backdown as a mobile platform for hand release of captured dolphins. The craft produces noise and bubbles, which may also act as a deterrent to dolphins swimming toward the vessel and swimming away from the release area during backdown. With no exposed propeller, the craft could quickly travel in and out of the net, passing over the corkline without risk to the dolphins of entanglement.

Since the demonstration, several vessels of the international fleet have tried the jet craft as a dolphin rescue platform. Early trips revealed mechanical and structural problems with the recreational craft due mainly to the rigorous conditions of commercial purse seining. Two units being tested by the U.S. fleet have undergone some structural and mechanical modifications by both the dealer or manufacturer and by the vessels using them. This has improved reliability but some problems remain. Increasing the crew's familiarity with the craft and storing a suitable inventory of spare parts should yield better reliability. An aluminum version or one patterned after a rigid-hulled inflatable also may increase the utility and durability in this application. A Mexican company is currently developing a jet boat for use in dolphin rescue (M. Hall, personal commun., 1990). In addition, several manufacturers of outboard motors produce propellerless jet models.

Current Profiler

The Doppler current profiler is a hull-mounted sensor that measures the speed and direction of currents at various depths below the surface. Commercial versions might be useful in providing subsurface current information to vessel operators before setting the seine. One product, designed for commercial fishing, provides subsurface current values at any selected depth (Summers, 1990). Another product, designed for oceanographic applications, provides a profile of the current over a continuous range of depths, plotting the results on a video screen (M. Hall, personal commun., 1990).

Such information could be valuable to the captain in determining whether subsurface currents are likely to cause a distorted set with collapses or canopies. If the current is not too strong, a set can be made with much less chance of a malfunction. The presence of a mild subsurface current, if its direction and speed are known ahead of time, does not necessarily spoil a properly oriented set. An excessive subsurface current would indicate the need to move from the area and fish elsewhere. With time, these devices may provide evidence for a correlation between tuna abundance and subsurface currents. Such an understanding could assist in locating tuna not associated with dolphins.

Pear-Shaped Snap Rings

An innovation that has been adopted by almost half of the U.S. fleet is the pear-shaped snap ring. Introduced primarily as a time-saver and a way to avoid the dangers of a conventional ring stripper, the snap ring helps to reduce the time before backdown begins. Instead of a simple steel ring, the new design is pear-shaped and has a spring-loaded gate portion that can be opened for inserting or removing the purse line. In conventional purse seining, once the rings are up, the rings must be transferred from being supported by the purse line to being supported by the cantilevered ring stripper. With the snap rings, this process is unnecessary and sacking up can begin immediately after the rings are up. This innovation can save up to 15 minutes, reducing the time that the dolphins are exposed to canopies or net collapse.

Small-Mesh Medina Panel

It has been suggested that placing extra slack in the large-mesh (4¼-inch) netting below the Medina panel would force the small-mesh (1¼-inch) netting above to open more during the backdown and make a better backdown. Thus, 10–15% more large-mesh netting would be distributed along the boundary with the small-mesh panel when the net is assembled.

Third Fine-Mesh Strip

The Porpoise Rescue Foundation is putting a third fine-mesh strip aboard two vessels. Adding this strip below the second fine-mesh strip may help save any dolphins that dive deeply as the backdown is being performed. This strip is made with 2-inch mesh. The 2-inch mesh is being used because it has less drag than finer meshes do while being pulled through the water during backdown.

Safety Crook

The safety crook is the same aluminum pole that is used in swimming pools to save swimmers in trouble. One end of this pole has a double open-end arc. During the backdown, this safety crook can be used in a speedboat to help guide the dolphins toward the opening that will let them free.

MAJOR MODIFICATIONS OF CURRENT METHODS

In the previous section, modifications and techniques were discussed that could be incorporated into present purse seines with little cost. In most cases, the potential benefit of the small modifications is the reduction in dolphin mortality during the backdown process. The basic freeing process remains unchanged. This section considers more sweeping changes in gear and methods. Most of the concepts presented have not been tested adequately to assess their impact on dolphin mortality and fishing effectiveness.

This section is organized around modifications to reduce two fundamental problems—canopies and roll-ups—and then continues discussion of four classifications of modifications—barriers, species selectivity, backdown-channel changes, and other purse-seine variations—that the committee believes show promise for reducing dolphin mortality.

The committee believes that the most promising major alterations in purse-seine gear are the following:

- Modifications in netting material.
- Modifications in hang-in ratio.
- Modifications to the purse cable.
- Development of lifting surfaces in critical parts of the net.

Modifications to Reduce Frequency of Canopies

Canopies are billows of netting along the perimeter of the pursed net and backdown channel; they are a common cause of dolphin mortality. These formations are caused by slackness in the upper portions of netting as the corkline is pulled sideways through the water. Causes of canopies can be subsurface currents, improper orientation of the backdown channel with respect to the wind, or improper control of the vessel during backdown. Although these causes cannot always be controlled, modifications to the purse seine that would make it resistant to canopies would be advantageous.

Canopies form when the downward pull on the net is overpowered by the drag forces due to water flow through the plane of netting. For most of the setting process, the weight of the rings, purse line, chain bridles, and leadline keeps the netting near the surface essentially vertical. As the net is pursed and the rings become supported by the purse line, only the weight of the netting itself remains to keep it hanging vertically below the corkline. The net configuration changes easily at this stage.

Some modifications to the purse seine singly or together would help prevent canopies. The most promising are the following:

- Netting of reduced hydrodynamic drag.
- Netting material with increased underwater weight.

The following also might help:

- A wider hang-in ratio (the ratio of the length of the corkline to the length of the webbing attached to it, typically around 80%).
- Netting twine of less stretch.
- Netting with additional weight in the form of a false leadline partway down the net.

These modifications are most useful when incorporated into the portions of the purse seine that become the backdown channel. Individually or together, these changes could reduce the tendency of the corkline to precede the netting when flow is normal to the netting panel. In general, a low-drag, fast-sinking net is advantageous in purse seining. With the possible exception of the wider hang-in ratio, these modifications should produce a more effective net to capture tuna.

At present, nylon netting is used for purse-seine construction because of its low cost, durability, flexibility, and sinking characteristics. Alternative netting materials such as Dacron* (polyester), Kevlar (aramid), and Spectra (ultra high molecular weight polyethylene) may offer advantages, but Kevlar and Spectra are more expensive than nylon. Dacron, in particular, has proved advantageous in other commercial fisheries. Because of its higher density, greater strength, and reduced stretch, its consideration in tuna purse seining may be appropriate.

One California netting company has built a purse seine in polyester, which, during use by one vessel, was found to be productive (H. Medina, personal commun., 1990). The company reports a weight advantage, cost advantage, and a sinking rate twice that of a nylon seine of equivalent strength (Kirkland, 1990). The reluctance of seiners to change netting material remains unexplained. Through attrition, the cost of such a change would be small since seine nets have a useful life of only 2–5 years (H. Medina, personal commun., 1990).

Modifications to Reduce Frequency of Roll-Ups

Roll-ups happen when the lower portion of the seine becomes twisted around the purse cable. Specific data on the frequency of occurrence of roll-ups are not available because roll-ups typically are categorized as a "major malfunction." In general, sets in which major malfunctions occur have nearly four times the normal rate of dolphin mortality (DeBeer, 1980).

In the following discussion of roll-ups, the related event of netting or bridles becoming snagged on the purse cable is also included. Roll-ups typically occur during the setting of the net when the cable is close to the netting. The bad effects of a roll-up begin during the pursing operation when a portion of the

---

* Dacron and Kevlar are trademarks of the DuPont Corporation. Spectra is a trademark of Allied Signal, Inc.

cable, entangled with netting, is pulled through the purse rings. The cable invariably becomes jammed and when it comes within reach of the crew, the hauling process must stop while the purse line is cleared. During this delay, net collapse and dolphin entanglement can occur.

The fishing industry has been strongly motivated to reduce the occurrence of roll-ups not only because of dolphin mortality but also because of the effort and danger involved in clearing roll-ups. Industry has eliminated some of the causes, such as reducing the occurrence of broken wire strands through proper maintenance of the purse cable. The use of longer chain bridles has helped to increase the space between the purse cable and the lower edge of the netting. However, these longer chains can themselves cause problems by wrapping around the purse line if they become slack.

In 1972, the Southwest Fisheries Center identified an important cause of roll-ups: the rotation of the purse cable during changes in its tension. Experiments with cables that were constructed to be torque balanced were begun aboard commercial tuna seiners. Though notable successes were experienced, cost and difficulty of splicing this unconventional wire rope prevented adoption. In addition, twist induced by the low-helix, three-strand torque-balanced cable passing over the purse block sheaves continued to cause occasional roll-ups.

Today, the delays during the hauling process remain an important cause of dolphin drownings. Despite some lingering problems, however, the torque-balanced cable can reduce significantly the frequency of roll-ups. Fortunately, alternatives to conventional wire rope are now available.

The successful introduction of Spectra braided line as a replacement for conventional wire rope in a variety of commercial fishing applications suggests its use as a purse cable. It has been reported that the Spectra line can last over 10 times as long as wire rope (Nye, 1990). Because of strength/diameter ratios similar to steel, few changes are needed to accommodate Spectra. Winch drum capacity would remain adequate. In addition to being torque-balanced, none of the problems associated with broken strands would exist, and splicing would be much easier.

The use of lead-core synthetic line as a replacement for chain along the lower edge of the seine may also be of value. This material was used by many of the San Diego boats in 1976, just after its introduction by Sampson Ropes (H. Medina, personal commun., 1990; T. Bergen, Sampson Ocean Systems, Inc., personal commun., 1990). Sink rates were adequate but its stiffness caused problems during stacking of the net on deck. The synthetic leadline was abandoned in favor of chain.

The stiffness of the leadline during initial use of the product was caused by the type and construction of the lead used in the core. The lead used had to be soft enough for extrusion and was notched to give it flexibility. However, under strain, this lead core would tend to deform, fusing the notches and causing stiffness. Recently, the manufacturer has introduced a harder lead, in cast ingot form, which has excellent flexibility that will not degrade under strain. Benefits that might accrue from the use of this new material would be reduced purse-line jamming, fewer roll-ups, faster sinking rates, and quieter setting while school fishing.

Modifications by Insertion of Barriers Between Tuna and Dolphins

One of the significant drawbacks among the present methods used to remove dolphins from purse seines is the loss of tuna. Increasing the opportunities for dolphins' escape also increases the risk of losing tuna. Certain techniques that might be appealing because of their possibilities in reducing dolphin mortality are unacceptable commercially. It is well established that tuna are far more likely than dolphins to escape from pursed nets.

With this in mind, NMFS's Southwest Fisheries Center experimented with a "backdown zipper" system that would separate the dolphins in the backdown channel from the tuna remaining in the main body of the net. Their approach was to use a line passing under the net partway up the backdown channel. During backdown and when the dolphins were beyond the line, the line would be pulled up, choking off the channel. With the two species separated, efforts to provide escape opportunities for the dolphins could be increased. Unfortunately, the dolphins were not always found on the right side of this zipper line, and the shallow channel would collapse. Experiments were discontinued. It is possible that attempts to learn how to maneuver dolphin herds inside the net would alleviate this problem.

The motivation for such gear modification remains sound and deserves continued thought. Opportunities may exist during the set when the two species are spatially separated. At the completion of pursing, dolphins are typically at the end of the net away from the seiner (Norris et al., 1978). During backdown, the dolphins are swept toward the apex of the channel. In both cases, placing a barrier between the species might be feasible. Deploying such a barrier is a challenge, because both species can react adversely to the disturbances caused by such activity.

Though still entirely conceptual, the design shown in Figures 7-3 through 7-5 can serve as an example of a barrier method. A netting panel of suitable size and shape can be buoyed into place at the proper time. Before deployment, the panel would be secured to the seine netting itself as shown in Figure 7-3. At the appropriate time, a release cord would then allow the panel's false corkline to pull the barrier to the surface, thereby separating the dolphins in their own netting compartment as shown in Figure 7-4. The main corkline and the netting below it would then be released and the dolphin herd would have an obvious escape route as shown in Figure 7-5.

The design and incorporation of this barrier net into the purse seine

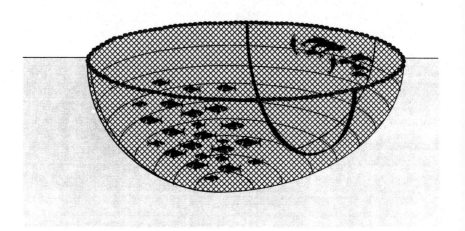

FIGURE 7-3 A full, deployed purse seine after pursing. The seiner would be to the left. Drawing by C. Goudey, Massachusetts Institute of Technology, Cambridge.

without excessively complicating the normal net-handling procedures would require careful thought. The principal advantages of this method are that (1) it could be done early in the set immediately after pursing, and (2) the backdown process would be eliminated.

Another approach is placing such a barrier net in the area of the backdown

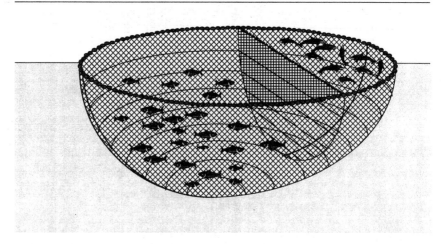

FIGURE 7-4 The barrier net has been released and has buoyed to the surface. Drawing by C. Goudey, Massachusetts Institute of Technology, Cambridge.

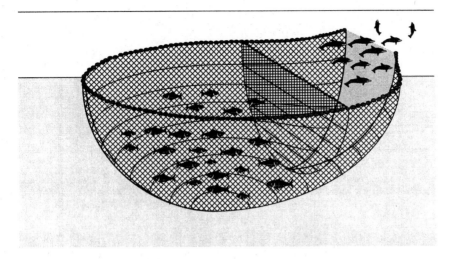

FIGURE 7-5 The main corkline and net are opened to release the dolphins. Drawing by C. Goudey, Massachusetts Institute of Technology, Cambridge.

channel to block the movement of tuna out of the main body of the net. With such a barrier in place, far more obvious escape opportunities could be provided for dolphins without risk of losing tuna.

Essential to the success of these methods would be an awareness of the movements of both species in the seine and knowing the right time to activate the raising of the barrier. As discussed in Chapter 5, our knowledge of the tuna-dolphin bond and the movements of both is far from complete. The rational development of such gear modifications awaits the development of a better understanding of species behavior.

Species Selectivity

The tuna-dolphin problem is a gear-selectivity problem. When netting is involved, the typical approach to such selectivity problems is the use of a specific mesh size that will retain certain fish based on size or herding traits. The situation is simpler when the by-catch species is smaller or is herded less easily than the target species. In such cases, a mesh size is used that retains the desired species while allowing the smaller or more elusive species to escape. The inverse problem—e.g., fish by-catch in a shrimp fishery—can also be solved using mesh size, but less directly. A typical approach is the use of a false net or barrier that deflects the larger or easily herded species away, while the smaller or less easily herded species passes through into the portion of the gear that ultimately retains it. The turtle excluder device is an example of a solution to the inverse problem.

Dolphins are far more easily retained in a net than tuna. The problem is therefore inverse: The dolphins must be separated from the tuna using a barrier through which the tuna will want to pass. Once the species are separated, further steps can be taken to release the dolphins, while precautions are taken to prevent tuna from recrossing the barrier. Implementing such a technique in a practical tuna purse-seining operation is far from trivial.

Information on tuna-dolphin behavior is inadequate for devising or implementing purse-seine modifications involving mesh selectivity as a means of mitigating the tuna-dolphin problem. More information is needed on the response of each species to various types of barriers. Many attempts to modify gear are based on behavioral differences between tuna and dolphins, as described in Chapters 5 and 6.

One possible approach would be to allow the tuna, which are more active in the net than dolphins, to escape the main body of the net into a secondary enclosure through a passage that discourages their return. The development of such a tuna "check valve" or "diode" has not been attempted, but there are many tools that could be brought to bear. Light, sound, mechanical flaps, or some combination of these could be incorporated into an opening, which would represent an inviting escape opportunity in one direction but prevent passage in the other.

Conceptually, a short tunnel with flexible reeds pointed in radially but biased in one direction would function to allow tuna passage only one way. A simple cone of netting as used in a fyke net or fish trap might also work. Determination of the size and clearance requirements of such a device would need some experimentation. Incorporating it into a commercial operation would require some ingenuity.

If, on the other hand, a directional light or sound source were discovered that repelled tuna, its placement around a hole in a panel of netting and pointed "downstream" might be an elegant approach. The principal appeal of this concept is that the dolphins would not pass through such an opening, and after the tuna had passed through, the main body of the purse seine could be opened and the dolphins could leave without running the backdown gauntlet.

The passage that the tuna would take would lead into a netting chamber that could be hauled aggressively as it would contain no dolphins. This chamber could take the form of a small pursed seine or simply a large sack of netting. Its design would require consideration of the quantities of tuna that would be encountered and their weight.

Modifications to Improve Escape of Dolphins from the Backdown Channel

The portion of the purse seine that forms the backdown channel is a relatively small part of the overall net; therefore, design modifications may be feasible for the channel that would be impractical for the entire seine. Changes or additions that add to the predictability or controllability of the backdown channel's geometry could assist in the reliable release of all dolphins.

Ideally, the backdown channel would remain open and functional, regardless of the behavior of the dolphins, the strength of the currents, the length of the release process, or the skill of the crew. The use of special components in the backdown channel that have controllable flotation, stiffness, or dimension but that still pass through the power block may have merit. Flexible hose that could be pressurized with air or water to provide near-rigid structural support may be of use. Control lines that change the shape of the netting or leadlines that are strategically placed might help form a more enduring backdown channel. Such an engineering solution is complicated by the poorly understood behavior of both dolphins and tuna. It is safe to say that few ideas representing truly novel approaches have been tried, much less properly evaluated in commercial conditions.

One concept that has been tested is the use of a backdown board to prevent the collapse of the channel. This board is a hydrodynamic lifting surface rigged and positioned to pull the corkline outward with the flow of water during backdown. The design was tested by IATTC, was found ineffectual, and no further experiments have been done (Bratten and Guillen, 1981). However, based on such limited experience, the concept of using hydrodynamic force to keep the backdown channel open cannot be dismissed.

The backdown board suffers from the fact that its performance depends on proper rigging and careful deployment. In addition, the device needs to be large—too large to be practical—to provide the force necessary. A more appealing alternative to the backdown board is a series of smaller flexible devices, attached to the corkline or the netting itself, that would remain in place on the seine. Flexibility and small size would be required since passage through the purse block is essential. Recent developments in headrope kites suggest that such flexible hydrodynamic lifting surfaces are compatible with the rigors of commercial fishing (Goudey, 1987).

Figure 7-6 shows a device that uses small annular lifting surfaces that would be attached to the portion of the corkline needing to be pulled out. Though flexible, these devices would retain their shape. Their size and spacing would be based on the lift forces required. They would be attached to the corkline so that they would align on the outside of the net, pulling outward because of backdown flow.

Figure 7-7 shows another device that could be attached to the panel of netting below the corkline. These small, flexible, kite-like panels would produce incremental amounts of lift, forcing the panel of netting outward. Both devices are conceptual and would require considerable development and testing before their value could be determined.

In addition to maintaining the integrity of the backdown channel, controlling the escape route at its apex would help to reduce dolphin mortality. Systems that include inflatable corklines and operable openings in the netting

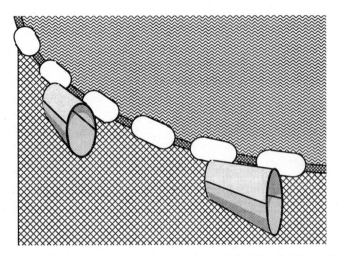

FIGURE 7-6 Annular lifting surfaces to maintain channel width during backdown. Drawing by C. Goudey, Massachusetts Institute of Technology, Cambridge.

have been proposed and attempted. A device called the "apex flapper" was tested with no success in 1977 (Coe et al., 1984). In related tests, a downhaul gate was rigged, but malfunctions prevented its evaluation. The limits of this approach have not been reached either in concept or in practice.

A U.S. patent has been issued on a modification to the corkline of a seine

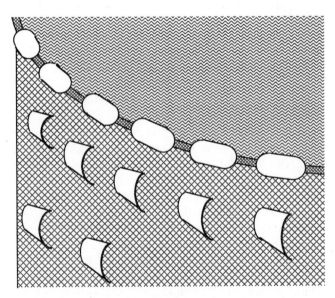

FIGURE 7-7 Netting panel kites for preventing channel collapse. Drawing by C. Goudey, Massachusetts Institute of Technology, Cambridge.

that could be used to allow the escape of dolphins while preventing the loss of tuna (McKnight, 1979). It incorporates hoses that would be selectively pressurized with air or water. This invention is not known to have been built or used in the fishery.

No Backdown

The backdown procedure is an effective way of getting most of the dolphins out of the seine most of the time. Since its introduction in the early 1970s, most of the effort spent on reducing dolphin mortality has concentrated on perfecting the backdown and the portions of the net that make up the backdown channel. However, the procedure is difficult and has opportunities for mishap—indeed, most dolphins die as the result of mishaps. Dolphin release may be accomplished in ways that do not require such precise coordination of seine handling, maneuvering, and the full cooperation of the weather.

One reason for an apparent preoccupation with backdown is that the purse seine will not remain open indefinitely. Unless its internal volume is continually decreased, it does not retain the desired hemispheric shape. Wind, waves, currents, and gravity conspire to collapse the seine and eventually folds and canopies of netting will begin to take their toll on the dolphins inside.

Intervention by the fishermen is necessary to prevent such net collapse. The sideways pull on the seine by the net skiff and the use of the speedboats to tow the seine back to its proper shape can delay collapse. There are reports of a seine being held open overnight by three speedboats while mechanical problems on the seiner were being repaired (D. Cormany, IATTC, personal commun., 1990). Weather conditions had to be favorable for this to be possible.

The backdown procedure does not normally require the intervention of speedboats. Sacking up the net and pulling in the corkline bunches keep an outward flow of water through most of the seine netting. Once a proper backdown begins, the netting remaining in the water tends to be held open by the flow.

If a simple means of maintaining the shape of a fully pursed seine were found, alternative techniques for the removal of dolphins could be developed and employed. Such techniques are currently impractical because there is not enough time to accomplish individual release of dolphins before collapse begins. However, the force required to maintain the shape of a seine must not be underestimated. Even the strong pull of the speedboats is insufficient when weather or current conditions are adverse.

The use of a flexible tubular device to replace or supplement the conventional corkline has been suggested as a way to produce a net perimeter with variable rigidity (D. Cormany, personal commun., 1990). Such a tube could be pressurized to prevent flexing of the tube and collapse of the seine. If the tube were inflated with air, the structure could actually replace the corks. If it were inflated with water, the corks would be retained and the tube would be hung parallel to the corkline. The diameter and pressure needed to achieve the required rigidity over a net perhaps 4,000 meters around have yet to be determined. Whether such a tube could be run through a power block is yet another issue.

Another approach would be developing a seine float that allows the corkline to flex in only one direction. If the perimeter of the net could take only a straight or convex shape, the inversions of the corkline associated with net collapse would be impossible. Passing such a corkline through the power block and stacking it on deck would require special handling techniques and might be impractical.

Other Purse-Seine Variations

Other methods have been proposed for separating tuna and dolphins. They range from bubble barriers to acoustic signals to helicopter-deployed extraction devices and have varying degrees of potential. However, few of these ideas have received more than a cursory examination.

POTENTIAL BEHAVIOR-BASED RELEASE OF DOLPHINS

Pre-Backdown Release of Dolphins

There is evidence that the "protective structure" of a dolphin group and the dolphins' unwillingness to cross a gate such as the one provided by the backdown channel are closely correlated. Dolphins could be released more efficiently from the net if such a correlation were understood better (Norris and Dohl, 1980b; Norris and Schilt, 1987).

A sharp behavioral difference exists between tuna and dolphins that might be used to design release methods. Tuna schools will flow through rather small openings in a net. Dolphins instead balk at passing through openings three times larger or more than a tuna school would use (Norris and Dohl, 1980b). H. Medina (personal commun., 1990) reports that tuna will sometimes flee the net through as slight an opening as a portion of the corkline that is locally depressed.

While we make no attempt here to design such a separation and release system, any such pre-backdown release method seems to require the following:

• The ability of fishermen to maneuver the trapped dolphin herd easily, which is apparently possible.

• A net-based device using the difference between the escape patterns of tuna and dolphins that could be deployed or incorporated in the net design within the larger seine and that could separate fish from dolphins.

• When fish and dolphins are separated, the lowering of a gate that will allow release of dolphins at some point on the net perimeter.

Devising such a system would require a joint effort by knowledgeable fishermen, behaviorists, and gear designers.

Because of the possibilities of this pre-backdown release method, it is worth examining the data on dolphin and tuna gate behavior. A gate of sorts is now used for dolphin release when the backdown channel is pulled hard by the seiner, depressing the corkline at the end of the channel 1–2 meters beneath the surface. The effect of this transitory opening is not to produce a gate through which dolphins will rush, but rather to provide an opening through which dolphins are tumbled willy-nilly, many moving out backwards, and sometimes swimming back into the net. The fishermen must carefully guard such a temporary gate because tuna will use it at once if it is available to them.

The remarkable reluctance of dolphins to pass through even large gates is well known by dolphin trainers (DeFran and Pryor, 1980). Gates that are 2 meters across and deep are regarded by captive dolphins as frightening barriers. Only after concerted training will individual dolphins rush through such gates. Even then such a passage always seems to involve an element of fear.

Recently, Norris et al. (in press) suggested that the central problem is that such gates do not accommodate the predator avoidance structure of the dolphin herd. The smallest protective dolphin herd is postulated to be approximately 6–10 animals arranged in a fluid geometry, the spacing of which is based on maintaining maximum visibility of the signals of neighbors and minimum reaction time to them. Such a protective group (called the basic school (herd) unit) with its geometry intact presumably cannot be accommodated in a small gate.

The first experiments related to this idea were those of Perrin and Hunter (1972), who attempted to define the gate size through which captive Hawaiian spinner dolphins could be induced to pass. These tests were inconclusive and seemed to predict a gate size much smaller than that used willingly by intact herds. The experiment did not include an intact basic school unit, and hence the reactions seen were not the reactions of a herd.

The second tests were the "hukilau tests" by Norris and Dohl (1980a), who attempted to learn the minimum barriers and dimensions of openings that would contain or release entire wild Hawaiian spinner dolphin herds. These dimensions probably bear a better relation to the dimensions of gates that dolphins will pass voluntarily through in purse seines than those defined by the Perrin and Hunter tests.

The important features of this test history can be summarized as follows:

- Much smaller gates will allow passage of a tuna school than will allow passage of a dolphin herd.
- The sinking and stacking of dolphins in the backdown channel may be due to their protective herd being compressed to such an extent that visual signaling between dolphins is no longer functional. The dolphins may be left without their protective geometry.
- The reluctance of dolphins to escape from the backdown channel on their own may also be due to the fact that the space provided is too small to allow the intact passage of the basic school unit.
- The predicted minimum gate for oceanic dolphins is about 7 meters wide and 7 meters deep. These dimensions correspond well to the gate size that allowed escape of spinner dolphins from the original hukilau tests (Norris and Dohl, 1980a). No predictions appear to exist for tuna escape.

Backdown Release of Dolphins

Much attention has been given over the years to refining the backdown-related seine-handling technique and the gear associated with it. In skilled hands, with good weather, and with smoothly working gear, it can be operated with very low mortality rates per set. Most mortalities occur during problem sets or are caused by inexperienced captains attempting the backdown method. Therefore, efforts to teach its intricacies to all the captains in the fleet are clearly the most promising route for immediate reduction of kill during seining.

Since fishermen are under constant pressure to produce short trips, they want to use every available working moment to catch fish. This incentive means that sets will be started when schools of fish are found, which may be late in the day, and may extend into dusk or even darkness. In such sets, mortality of dolphins rises markedly; however, sundown sets are prohibited by NMFS regulations as well as those of other countries. Usually, this elevated mortality is blamed on the increased difficulty of net handling and of carrying out seamanship in low light levels. However, the dolphins in the net may experience even greater difficulties than the fishermen. At dusk their ability to avoid entrapment or to assist in their own release may be severely reduced, because the reduction in illumination beneath the sea surface during dusk is much more profound than in air (100 times or more) (McFarland and Munz, 1975).

Thus, in low light levels, dolphins may have sharply increased difficulties in responding to each other and may provide even less assistance to those trying to release them than they do during daytime sets. If sundown sets come to be practiced again, underwater illumination of the net in the region of the apron should be tested.

Dolphin Behavior on Release

Observers at the lip of the backdown channel during the backdown operation report seeing almost passive masses of dolphins being sluiced toward them as the ship pulls the net (Norris et al., 1978). The backdown channel may be pulled 1–2 meters below the surface at this time, and dolphins may slide out forward or backward. The instant a dolphin perceives that it is free of the net, it is almost immediately galvanized into action and streaks away from the seiner and its net. This almost instantaneous change in behavior—from total passivity to high-level activity—is remarkable (Pryor and Kang, 1980) and makes it clear that the dolphins in the net are not in a catatonic state in which they cannot protect themselves at all. Given an opportunity (such as having the channel kept open or preventing canopies), they will rise to the surface to breathe. It is possible that research into further exploitation of this readiness to escape could lead to further reduction of dolphin mortality.

BREAKING THE TUNA-DOLPHIN BOND

Any program that seeks to reduce dolphin mortality in the ETP yellowfin tuna fishery by understanding and subsequently modifying the behavior of tuna or dolphins must be based on the answers to a number of fundamental questions.

- What is the nature of the tuna-dolphin bond?
- What regulates the persistence of the bond?
- What underlies the different responses of tuna and dolphins to seining?
- Why is it that tuna are escape artists once they are encircled in the seine, and dolphins refuse to help themselves and must be assisted out of the net? This difference lies at the heart of the problem of dolphin release and may be due to the different senses used by dolphins and by fish to regulate the structure of their schools—vision and hearing are used by dolphins, and vision and mechanoreception (through lateral lines and head canals) are used by the fish.
- What underlies the different responses of various species to each other and to floating objects? Clearly, the various tuna species differ greatly in size and other properties, which affects whether they associate with dolphin species. If we understood the underlying causes, the nature of the bond itself might become clearer. The association is probably food-based, and tuna probably follow dolphins, although some believe that dolphins follow tuna. Fish-aggregating devices usually gather smaller tuna than those that collect with spotted dolphins. One center of aggregation is drifting in the water and the other is swimming. Does motion or lack of it regulate associations?
- What oceanographic variables underlie the associations between tuna

and dolphins? Analysis of oceanographic and seasonal catch data, especially current, trophic, and thermal structural data, should allow assessment of the correlates of aggregation, provided such catch data can be freely obtained.

• What are the effects of seining upon dolphin societies and population structures?

• What are the specifications of a net opening that will allow dolphins to swim from the net on their own volition? A research group has begun to define the spatial relations that operate within dolphin herds, and these ideas in turn may help define the opening in a tuna seine that will allow dolphins to escape on their own. The theory says that a basic protective school unit exists that dolphins will not abandon and that it must be released as a unit. This unit plus undefined space on either side of a dolphin in a herd may define minimum openings needed for quick release of dolphins.

ALTERNATIVE METHODS OF LOCATING YELLOWFIN TUNA

Because tuna fishermen find tuna by looking for birds and dolphins, an obvious way to prevent dolphin mortality is to find ways to locate and catch tuna when they are not associated with dolphins. In all the world's oceans, other methods are used. Even in the eastern Pacific, much small yellowfin tuna is found and caught in ways other than rounding up dolphin herds. The question of how to catch the large yellowfin that usually are caught with dolphins comes up in every meeting and workshop on the tuna-dolphin problem (Ralston, 1977; Hofman, 1981; DeMaster, 1989). Variants of all the methods for locating large tuna proposed here have been discussed at those workshops. Researchers propose ideas that usually languish because of lack of funding for the research that needs to be done to test them.

Acoustical Methods

Large schools of any fish moving in the water must make noise. If we knew the characteristics of the sound produced by tuna, then listening devices could tell fishermen where the fish were. One such device is a towed array, a streamer of hydrophones that is towed behind a vessel. It is sensitive only to sounds that come from the sides, which eliminates noise made by the towing vessel and facilitates locating and tracking sound sources. Towed arrays are used by the geophysics industry for petroleum exploration and by the Navy for finding and tracking submarines. Thomas and Evans (1982) used a towed array aboard a tuna seiner to evaluate its usefulness for detecting dolphins and tuna. Sound spectra taken when large numbers of yellowfin tuna were present showed a consistent increase in broadband acoustical energy between 1 and 6 kilohertz. Detecting the signal in their tests required a trained technician to operate a spectrum analyzer and to maintain the array and its electronics.

Because the fishing effort concentrated mostly on dolphins, the array's potential for locating tuna apart from dolphins remains untested. However, the subtlety of the tuna's sound implies that towed arrays probably would be unable to detect tuna more than a mile or two away from the vessel unless they become more effective than those available in the early 1980s. Towed arrays are likely to be impractical for use in the tuna fishery, at least as a primary tool.

Listening devices may be valuable with fish-aggregating devices (FADs). A hydrophone under each FAD would pick up the sounds in its vicinity. Digital circuitry could then check the characteristics of the sound and store information about the presence of tuna for retrieval by a seiner. One attempt was made to equip logs with sonobuoys, but no tuna were located (F. Awbrey, San Diego State University, personal commun., 1990).

Another acoustical method for locating tuna is sonar, which is used very successfully in the western Pacific tuna fishery, as well as in fisheries for other species. Fishermen in the eastern Pacific avoid using sonar because dolphins detect sonar at long range and move away from it, taking the tuna with them. A solution may be to use only frequencies above the hearing range of dolphins, about 150 kilohertz for one species and unknown for others. The major disadvantage of high frequencies is limited range—the higher the frequency, the shorter the range. However, high frequencies give better resolution and thus would enable surer identification of species and their numbers. Properly used, a high-frequency scanning sonar may let fishermen follow tuna and dolphins and set their nets whenever the fish move away from the mammals. If the fish return to the dolphins as soon as a disturbance starts, this technique would not work, but it ought to be tried. By not using sonar, fishermen also may be passing tuna schools that are below the surface but within reach of their nets.

The use of such equipment, in addition to or instead of helicopters, to determine the number of tuna in a school also might help eliminate sets on dolphins when only a few fish are present.

Other Methods

Remote sensing satellites now send back information about oceanic conditions all over the world. Timely information on sea-surface temperature and water color is available to everyone at a reasonable cost. Southern California albacore fishermen use water-color data to find the edges of clear water and green water where they catch albacore.

Satellite imagery may be of use in locating tuna not associated with dolphins. Given surface color and temperature, current boundaries and areas of high and low mixing can be identified. With this information, any areas with non-dolphin-associated tuna could be discovered. Such information could hold an important key to developing a dolphin-safe fishery for tuna.

Shipboard receivers can be purchased that receive images directly from satellites or from shore-based transmitters, much like a weather facsimile, and some are being tested at sea by tuna fishermen. Interpretation of the information is required, and its usefulness will depend on the expertise of the captain.

Human observers on ships or in aircraft need tools to aid their limited ability to see what is under the surface. A relatively new technology involving the use of LIDAR (light-induced detecting and ranging) has been proposed for use in tuna fishing to detect subsurface schools of fish (Summers, 1990). If this laser-based technique is successful, it would help to detect tuna schools without reliance on associated dolphins. LIDAR can detect profiles of oceanographic conditions to at least 30 meters depth. These devices provide detailed information about such things as phytoplankton biomass and scattering layers (Hoge et al., 1988) and are said to be useful in the menhaden fishery (B.D. Treadwell, Remote Sensing Industries, personal commun., 1990). The manufacturer is working with Honor Marine Communications of San Diego and some vessel owners to test whether one of these devices, adapted to work from a small helicopter, can detect tuna schools in the western Pacific (R. McCloskey, Honor Marine Company, personal commun., 1990). LIDAR and high-resolution sonar may be useful for studying the conditions where the tuna-dolphin bond breaks up, such as studying tuna at night to determine whether they separate from dolphins and become vulnerable to seiners. They also may be a good tool to use in deciding where to place FADs to maximize the chances of aggregating large tuna.

Another recent development that would be of value in searching for school fish is the synthetic aperture radar (SAR). Of particular interest to the military, this technology has proved capable of detecting the wave variations in the path of a ship long after it has passed. The wave-detecting ability of SAR could be used to detect "breezers" (tuna feeding on the surface causing local waves to form). SAR could locate the higher-frequency breezer waves from within ocean swells or even superimposed on ambient wind waves. The question is would high-resolution SAR in a helicopter detect surface activity, such as breezers, that human eyes miss because of the sun's angle or wave action?

Height is important for SAR to be most effective. Optimal angles are 30° to 40° to the surface. Therefore, ship-borne SAR would be much less effective than helicopter-borne SAR. An intriguing approach would be the use of a balloon to support the unit, increasing the search radius far beyond present visual range.

ALTERNATIVES TO DOLPHIN-ASSOCIATED FISHING

Fish-Aggregating Devices

Tuna fishermen call flotsam, such as trees and other debris, logs. Anchored or drifting artificial objects deliberately placed to attract tuna are known as FADs. Discussions about alternatives to dolphin-associated fishing for yellowfin tuna usually include using such floating objects to aggregate fish. In all the world's tuna fisheries, logs and FADs are important aids for fishermen (fishermen generally refer to all of them as "logs"). Purse seiners catch substantial amounts of tuna every year by setting nets on floating objects, but no one knows why tuna associate with floating objects or how strong or long-lasting the attraction is.

History and Perspective

Before World War II, Philippine fishermen started using anchored FADs, which they call "payaos," to attract pelagic fish. The successors of these FADs, along with natural flotsam, are now important fishing tools. In the western Pacific Ocean and the Indian Ocean, where tuna and dolphins appear to associate less frequently than in the ETP, purse seiners concentrate on free-swimming schools and flotsam-associated fish, and they rely heavily on logs and FADs for finding and catching tuna. In the right circumstances, flotsam and free FADs have some very advantageous features. Flotsam-associated fish are much less likely to leave the net during pursing than free-swimming schools are. Catches can be large but are on average smaller than the catches in dolphin sets. Sometimes a seiner returns to a log periodically for several days, taking substantial amounts of fish nearly every day and saving fuel in the bargain. FADs with transmitters are easy to find. If equipped with proper sensors, FADs can provide information about fishing conditions through radio signal, saving time and fuel and increasing catch per unit effort.

The success of FADs elsewhere in the world and the need to find alternatives to dolphins for aggregating tuna should make FADs an attractive option for fishermen in the eastern Pacific. Typically, however, in the Indian Ocean, school fish and flotsam-associated fish tend to be mixtures of skipjack and small yellowfin. Free-swimming schools are mostly yellowfin, and flotsam-associated schools are mostly skipjack (Table 7-1). FADs are expensive to place and maintain and they are subject to piracy. Boat owners, understandably, do not want to see their investment appropriated by someone else. There is also the question of who owns the fish under a FAD. Technological and legal steps may reduce those problems.

The experience with flotsam in the ETP has been that, in most areas, it does not attract the large yellowfin that associate with dolphins. Figure 7–8 shows

TABLE 7-1 Species Composition of the Tuna Catch in the Indian Ocean, January 1982 to December 1984[a]

Species	Flotsam		Schools	
	Metric Tons	Percent	Metric Tons	Percent
Yellowfin	10,356	24.4	39,855	78.2
Skipjack	30,031	70.7	10,132	19.9
Bigeye	2,075	4.9	725	1.4
Albacore	6	<0.1	292	0.6
Total	42,468	100.0	50,994	100.0

[a]Data from Hallier, 1985.

that landed fish from the nearly pure schools of yellowfin tuna caught in sets on dolphins in the ETP are large (>80 cm or 23 lb). In contrast, yellowfin caught in sets on logs are usually small (<80 cm, <23 lb) and they are mixed with larger amounts of skipjack (Greenblatt, 1979; IATTC, 1989b). From the point of view of tuna conservation and production, shifting the fishery toward logs could have several undesirable outcomes and is probably not ecologically sound. The fundamental question, therefore, is whether it is possible to develop FADs capable of attracting large yellowfin. Although tuna seiners in the ETP take large amounts of school fish and flotsam-associated fish, including skipjack and other species, yellowfin are the most important component of the catch in the ETP purse-seine fishery, comprising up to 57% of the catch between 1980 and 1987 (IATTC, 1989a). Of all yellowfin caught in the ETP, about 65% are caught on dolphins (IATTC, unpublished data). Overall, in the major oceanic fisheries, yellowfin comprised about 32% of the catch between 1980 and 1987 (IATTC, 1989a).

Status

As part of the effort by IATTC to evaluate alternative fishing methods, Guillen and Bratten (1981) anchored five wooden FADs in the ETP in water 1,700–2,100 fathoms (3,108–3,840 meters) deep, 420–600 miles offshore, between 9° N and 15° N. As usual for that area, the tuna they attracted were skipjack and small yellowfin, not the large yellowfin that are important in the dolphin-based fishery. Because of the small size of the tuna caught, and the high costs of deployment and difficulty of maintenance, IATTC concluded that this approach was unlikely to succeed. However, IATTC suggested that free-drifting FADs might be better, especially if they were instrumented so that seiners could monitor them for position and information about the fish associated with them, but IATTC still emphasized the small size of the fish (IATTC, 1989e) and the possible consequences of shifting to that resource. IATTC argued that as the fishery is constituted now, switching from dolphins

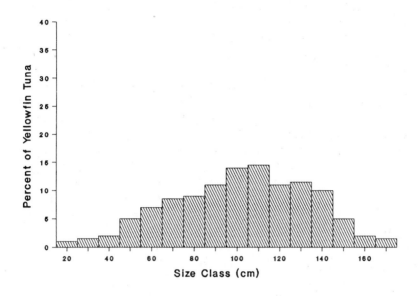

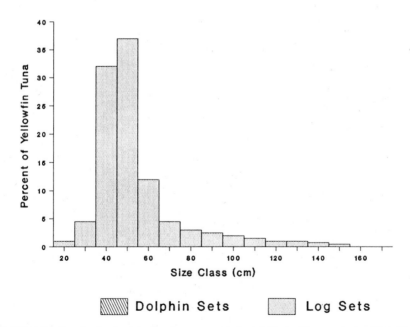

FIGURE 7-8 Size distribution of yellowfin tuna caught in dolphin and log sets during 1974–1985 in the eastern Pacific. Data from Figure 41, IATTC, 1989b.

to logs or drifting FADs could reduce effective recruitment as well as yield per recruit. Heavier exploitation of flotsam-associated fish in the usual places would catch a larger proportion of yellowfin while they were sexually immature and still below optimal size for maximizing yield per recruit (IATTC, 1989c).

The conventional wisdom that logs and FADs will catch only small yellowfin in the ETP may not be as well founded as it seems. The geographic distribution of log fishing in the ETP strongly reflects the sources of flotsam in south and central America. The areas of most intense log fishing are close to the coast, where logs are most plentiful but where the yellowfin tuna are young and small. As the logs drift westward in the currents on each side of the equator, they tend to become waterlogged and sink. Few natural logs survive in the ETP west of 120° longitude, yet the sparse data about the logs that are there suggest that FADs placed there would accumulate a larger yellowfin on the average than those found with logs in the usual places nearer to land (M. Hall, personal commun., 1990). In its program on alternative fishing methods, IATTC has preliminary data for the years 1970–1988 (Figure 7-9) showing that in some 10-degree squares, some yellowfin caught under logs are large (>100 cm). Although sample sizes are small, these observations suggest that experiments with FADs should not be done in the traditional log-fishing areas, but rather in locations farther from land where FADs have greater potential for aggregating large yellowfin. Support for this suggestion comes from the western Pacific, where large yellowfin tuna (>80 cm) often associate with flotsam. Fishermen say they catch more small yellowfin near shore and find larger fish by searching farther out. Placing FADs where larger fish are more likely to associate with them makes more sense than putting them elsewhere. A FAD-deployment experiment began in July 1991 conducted by IATTC, NMFS, and Mexico and supported by Bumble Bee Seafoods.

Potential

The chance of success in the endeavor to attract tuna away from dolphins or to develop an alternative aggregator of large yellowfin tuna would be much higher if we understood why tuna associate with dolphins and with flotsam. So far, however, we can do little more than speculate. To generate hypotheses, we need fundamental environmental and biological information that is still missing. Conjectures about the tuna-dolphin bond are that its basis may be location of food, protection from predators, active movement, environmental conditions, or a host of other factors. The correct hypothesis must explain why the bond between the same species of dolphins and tuna is more frequent or longer-lasting in the ETP than in other ocean areas where these species also are found together. It also must explain why large yellowfin aggregate with logs in some places and not in others.

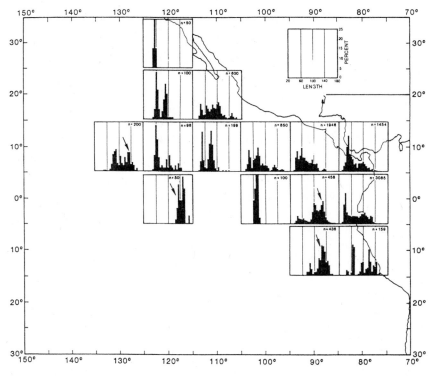

FIGURE 7-9 Size distribution of flotsam-associated pure schools of yellowfin tuna in 10-degree squares in the eastern Pacific. Squares marked with arrows have size distribution of flotsam-associated catch very similar to that of dolphin-associated catch in the same square. Source: Adapted from IATTC, 1991a.

Conclusions

The possibility of substituting FADs for dolphins as a means of aggregating large yellowfin tuna in the ETP has yet to receive the attention it deserves. Research has been going on for almost 4 years but has been underfunded. Essential information for establishing hypotheses has been lacking for far too long. Well-planned and funded research on why tuna associate with floating objects and dolphins deserves a high priority. Parallel programs also are essential to determine the properties of the best FADs, what technological devices can make them more efficient to use, and where they can be placed to attract the class of yellowfin tuna that are now caught mostly with dolphins. An institutional workshop has been organized by IATTC with industry support to study the association of tuna with floating objects in all oceans of the world that may provide answers to some of these questions.

Non-Purse-Seine Techniques

Alternatives to using purse seines to fish for tuna in the ETP exist. Government and industry workshops have explored this approach (DeMaster, 1989). A variety of techniques are used worldwide to exploit tuna and other midwater schooling fish, but few techniques offer the production rates now enjoyed by the modern tuna purse seiner and thus cannot compete economically with purse seiners at present. In addition, other fishing techniques are known to kill some dolphins (Northridge, 1984, 1991).

Bait Boats

The use of bait boats is clearly feasible but unlikely to be economically viable. As explained in Chapter 1, this was the method of choice before the introduction of purse seining. The use of hand-pole fishing is largely incompatible with the vessels currently employed in the ETP. Tuna-seiner designers quickly abandoned the low stern freeboard essential to pole fishing. Typical production rates of bait boats compare poorly with the rates averaged by today's purse-seine fleet. In addition, because of the labor-intensive nature of such a fishery, it is unlikely that the price offered for tuna by canners could support a U.S.-based bait-boat operation.

Longline Fishing

Longline fishing is a common method of fish capture and is the typical method used for tuna species, including yellowfin, in parts of the Atlantic, Pacific, and Indian oceans (see Chapter 3). Hook size and type of bait are specified according to the size and species of the targeted fish and, as a result, longline fishing can be quite selective.

An important consideration in comparing longline fishing with present techniques of purse seining used in the ETP is the differences in production rates and markets. The actual hauling process is time consuming, not to mention the time spent setting the longline, returning to the beginning of the line, and possibly waiting for bait to be taken. The productivity of longline fishing does not compete boat for boat with ETP purse seining. As a result, the use of the present tuna seiner as a conventional longline vessel would not be economically feasible.

Two variations of longlining may deserve further consideration. The first is the conversion of a seiner into a mother vessel for a fleet of suitably sized but launchable catcher boats. The present ability of seiners to launch and retrieve the large net skiff would need some modification to handle several smaller longline boats. In such an operation, the catcher boats would return frequently to the mother vessel for off-loading. All catcher boats would be taken on board during bad weather and during return to port.

The details of such an operation would need serious thought including economic assessment and optimization. The advantage is that the catching efficiency of a compact longliner can be combined with the endurance, searching ability, and capacity of the present seiner.

Another even more innovative approach incorporates the use of the seiner's helicopter to place a series of longlines in the path of a sighted school of tuna or mixed school of dolphins and tuna. According to the helicopter pilot who devised this technique, the longline could be placed without scattering the school, and the dolphins would not take a hook (K.R. Thomas, Thomas Atmospherics, personal commun., 1990). Because this operation would be directed at specific schools of tuna, catch rates per hook might be extremely high. Conversion costs for such an operation would be low. Experiments are needed to determine whether the tuna will stop to feed if their prime intent is keeping pace with the dolphins and to be sure that the dolphins will not take the bait.

The airborne aspects of this technique are the subject of a patent application. According to the inventor, conversion costs for such an operation would be low and negotiations are under way with hook, line, and baiting machine manufacturers with the intent of rapidly commercializing the method. An economic assessment of this technique would also be essential.

Midwater Trawls

Few alternative methods of fishing offer the productivity rates enjoyed by tuna purse seining in the ETP. A significant exception is trawling and a notable example is the U.S. pollock fishery off Alaska where 100-ton tows are common.

Tremendous progress has been made over the last decade in the design of gear for such high-volume trawling. High horsepower vessels, new materials, and the exploitation of the herding characteristics of pollock have allowed the development of midwater trawls in which the area of the mouth of the trawl is measured in acres. The applicability of trawling to tuna fishing in the ETP is unclear. Factors such as the herding characteristics of tuna, the towing speed required, the effect of vessel noise, and the presence of dolphins all prevent the direct extrapolation of most trawling experience.

Pair Trawls Several of the committee members believe that pair trawling is not a viable option for catching yellowfin in the ETP, but a few members felt it to be worth considering. A brief description of the French experience is thus provided.

In France, midwater trawls have been somewhat successful in capturing albacore in the Bay of Biscay (Prado, 1988). This technique was developed to increase the productivity of small (17–20 meters) longline vessels that normally have low catch rates. Pairs of these vessels were fitted with a midwater trawl normally used for bass. These 400–500 horsepower vessels were able to tow the net at 3.5 to 4.0 knots while achieving a vertical gap of 30–40 meters and a horizontal opening of 60-80 meters. Trawling was done at night on the surface of the water or close to it. The thermocline in this location is 10–15 meters, which is shallow.

The sound of the pair trawlers, the tow cables, and the 18-meter meshes in the forepart of the trawl were found to be effective in herding the tuna toward capture. Catch rates of 3–8 metric tons per day have been reported. Catch rates of purse seiners are considerably higher.

Incidental Catch of Dolphins in Trawls   In evaluating the utility of trawling as a substitute for purse seining, the incidental catch of dolphins in established trawl fisheries is relevant. As discussed earlier, little useful information is available to aid in the design of gear that would select only tuna from a mixed school of tuna and dolphins.

Anecdotal reports of bottlenose dolphins being caught in large midwater trawls used along the U.S. east coast exist. Reports compiled by NMFS on marine mammal mortality state a trawl-related incidental catch of 20 and 42 dolphins of all species for years 1987 and 1988, respectively (Waring et al., 1990). Clearly, trawls are not designed to spare dolphins. Specific techniques may be applied, however, to avoid their capture.

A Hypothetical Trawling Technique for Tuna in the ETP   The committee did not agree on the promise of this option; several members believe that it is not worth pursuing. Perhaps the greatest disadvantage is that if any dolphins were caught in such a net, they would quickly drown and would not be discovered until long afterwards. However, a few members felt that this option might be developed effectively, and therefore it is briefly described.

Conventional single-boat midwater trawling probably would not be useful in tuna fishing because of vessel noise and especially because of the high swimming speeds of tuna. By comparison with single-boat trawling, pair trawling offers the following potential advantages:

* Vessel noise may herd the tuna into the path of the net.
* Trawl doors are not required.
* Extreme spreads are possible.
* The trawl could be larger than a single-boat trawl because twice the horsepower would be available.

Pair trawling would not necessarily require purpose-built vessels. Adaptation of purse seiners currently in use in the ETP is feasible. Their seining capabilities may be retained, yielding a combination trawler-seiner. In addition, although present methods of tuna location could be retained, dolphins would not be encircled by a net and their herds would remain intact.

The combined horsepower of two 3,800-horsepower seiners would allow the use of a net with approximately 266,000 pounds of drag. Though the drag would vary considerably with speed, 7,600 horsepower would be sufficient to pull a very large net. Depending on the extent to which large mesh or ropes could be used, vertical openings of 550 feet and horizontal openings of 3,000 feet might be possible.

Based on the French experience, the towing speed does not have to equal or surpass the swimming speed of tuna, at least not at night. If the passage to the cod end of the net is viewed by the tuna as a possible escape route, high towing speeds may not be required. The same speedboats that are used in purse seining may be useful in tuna midwater trawling. Several of the small craft could be used to direct a school of dolphins into the gap between two approaching pair trawlers. Eventually, optimal values of pair-trawl speed and vessel separation would be determined.

Data on the vertical separation of tuna and dolphins during normal swimming and during a chase would be essential to allow the selective catch of tuna; indeed, if tuna and dolphins are not adequately separated, this technique might kill more dolphins than purse seining does. The depth of the headrope could be set either by varying the speed or the length of tow cable (typical of most midwater trawling) or by the use of surface buoys on pendants of specific length. The intention would be to place the headrope of the trawl between the two species or slightly deeper. Experimentation would be required to learn the best depth for minimizing tuna loss without risk of capturing dolphins.

Such a pair-trawling operation can be undertaken only if the vertical separation of the two species is predictable or if they do not associate at night. Certainly the depth of the headrope should be well below the deepest dolphin, even though the upper portions of the tuna school may be lost.

In addition to the risk of killing dolphins, a disadvantage of this method is the requirement of two seiners for one fishing operation. This problem is lessened, however, because most of the trip time is spent searching, and two vessels would offer increased search coverage.

Gillnets

Gillnets can be a selective method of capture in some fisheries. In other fisheries, or when used irresponsibly, they can be indiscriminate in their catch. Productivity with gillnets is similar to longline rates and falls far short of purse-seine rates.

Marine mammals often are entangled in gillnets, and it is extremely doubtful that a pure catch of tuna could be pulled from an aggregation of mixed species. A dolphin mortality of three animals per net has been observed in the Japanese driftnet fishery for albacore (Anon., 1990). The committee recommends against the use of gillnets as an alternative to purse seining in the ETP.

REGULATORY ALTERNATIVES

The behavior of the tuna fishermen can be changed voluntarily or through regulation. The tuna fishermen themselves have instituted or suggested changes in their tuna-seining processes over the past two decades. Some of these changes have been voluntary, and some have been stimulated by laws and regulations. Unless additional voluntary change is anticipated, further development of law and policy must be considered if behavior is to change and if dolphin mortality is to be reduced.

Any of the changes referred to in the above sections on alternative fishing methods, if not voluntarily adopted, would have to be the subject of regulatory action. Education and training, discussed in the next section, may stimulate voluntary changes in methods of tuna seining. The purpose of this section is to enumerate the major alternatives for regulatory action, the assumption being that voluntary action may not be sufficient to accomplish various public-policy goals with respect to the tuna-dolphin issue. The following discussion of the alternatives should not imply that any particular alternative should or should not be implemented. It is intended simply to show the range of possible public-policy options. The success of such options would depend on the degree to which they were implemented by all nations with boats fishing for tuna in association with dolphins.

The first set of options centers on regulatory alternatives that would further prohibit, directly or indirectly, dolphin mortality.

Immediate Prohibition of All Dolphin Mortality from Fishing

This alternative would consist of an immediate prohibition of purse seining on tuna associated with dolphins. The IATTC staff (J. Joseph, personal commun., 1990) is quite confident that a prohibition of dolphin fishing in the ETP would result in a substantial reduction in the overall production of yellowfin. In addition, if some of the fishing capacity were redirected toward smaller yellowfin, the overall biomass of yellowfin might be significantly reduced as well. Some of this loss would be made up by increased catches of skipjack, but the level of that catch cannot be predicted, because the availability of skipjack in the ETP varies greatly from year to year.

The IATTC staff is not able to make accurate predictions of the long-term effects of catching only small fish on the spawning biomass and subsequent recruitment of yellowfin because no apparent relationship has been observed between spawning biomass, as currently estimated (see Figure 7-10), and recruitment. This may be due to the lack of a relationship over the range of observed stock sizes or due to deficiencies in the data. IATTC is continuing to work on this problem. The lowest spawning biomass was estimated at about 90,700 metric tons (Figure 7-10). Females do not reach sexual maturity, on the

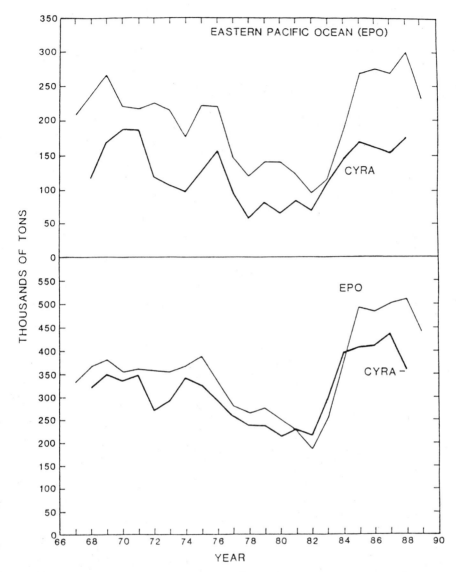

FIGURE 7-10 Estimates (in short tons) of the biomass of large yellowfin (upper panel) and of all yellowfin (lower panel) in the eastern Pacific. (1 short ton = 0.907 metric ton.) Source: Adapted from IATTC, 1990a.

average, until they exceed 40 pounds, which is about the average size of the yellowfin taken in the fishery for fish associated with dolphins (Figure 7–11).

The most recent IATTC estimate (IATTC, 1991b) of the long-term effect of a prohibition of dolphin fishing in the ETP is that it would result in a

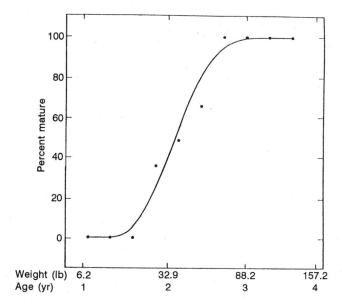

FIGURE 7-11 Percent maturity of female yellowfin tuna as a function of average weight and age. Data from IATTC, 1990b.

30–60% reduction in ETP yellowfin catch, as well as a significant reduction in yellowfin stocks, if no alternative method were developed to capture large yellowfin.

Dolphin Mortality Certificates

Under this alternative, certificates would be issued to the captain of each vessel in the tuna fleet for a certain share of the total allowable dolphin mortality. The certificates would be broken down into units of, say, 1–10 dolphins each. The permitted level of dolphin mortality for each captain would be equal to his number of certificates. These certificates could be nontransferable, in which case a captain would be limited to the mortality on his original certificates, or transferable among captains, perhaps on a market basis, in which case a captain could buy certificates from other captains. In either case, when a captain reached the mortality totaling the number of certificates he possessed, he would have to stop fishing on dolphin-associated tuna.

This alternative would make sense only if the total dolphin mortality would be reduced steadily over time. To achieve reduction in total mortality, each captain would have to surrender a percentage of his certificates each year. Under a transferable system, a captain could purchase more certificates from another captain. The price of a certificate could be determined by the market for the certificates among the captains, not by the government. Under this plan, the government would simply record certificate transfers and monitor compliance. Under a nontransferable system, all captains would reduce their dolphin mortality by an equal percentage each year.

The second set of options discussed below centers on alternatives that would create incentives for behavior that reduces dolphin mortality, as opposed to direct or indirect prohibition on dolphin mortality itself.

Incentives for Tuna Fishing with Alternative Gear

Under this alternative, positive incentives for switching to alternative fishing gears would be created, probably in the form of subsidies for purchasing or constructing vessels or gear designed for fishing methods other than purse seining.

Price Incentives for Fishing on Non-Dolphin-Associated Tuna

The argument has been made that one of the factors associated with dolphin mortality is the premium price paid for larger yellowfin tuna, which are more likely to associate with dolphins. This alternative would involve altering the price structure for tuna to eliminate the perceived or real advantages of purchasing larger yellowfin. This change would probably have to be made through the mechanism of a tax or tariff on yellowfin over a certain size or through a price incentive for other species such as skipjack.

The major canning firms in the U.S. market recently (1990) created such a disincentive by announcing that they would no longer purchase tuna taken in association with dolphins in the ETP. Although the mechanisms of this proposal are unclear, this move would appear to place a significant constraint on fisheries involving dolphin mortality, at least those delivering tuna to canneries owned by these firms. The important questions concerning this proposal from the point of view of this committee are the following:

• How will the proposed program be monitored? The initial understanding is that the existing U.S. and international observer programs will be continued, although they would have to be expanded considerably to serve a certification function for all countries and vessels involved in dolphin-related tuna fisheries. Will the monitoring and enforcement of this program be done through government programs, such as the NMFS observer program, through international or regional programs involving such organizations as IATTC, or through the mechanisms described in the Dolphin Protection Consumer Act of 1990 (P.L. 101–267, Title IX)?

• How will any potential embargoes on tuna products coming into the United States under the 1988 amendments to the Marine Mammal Protection

Act of 1972 (MMPA) affect international or foreign nation observer programs? What, in turn, will be the effect of changes in foreign-nation observer programs on the ability of canning companies or the U.S. government to monitor catches of tuna associated with dolphin mortality?

• What will be the effect of the canning-company decisions on the U.S. high-seas fleet operating in the ETP? Five general alternatives present themselves:

1. Fishing on school fish not associated with dolphins outside the 200-mile Exclusive Economic Zones of the Resource Adjacent Nations in the ETP.

2. Making arrangements with Resource Adjacent Nations to fish on school and log fish within their Exclusive Economic Zones.

3. Moving from the ETP to other fishing areas and fishing on school and log fish. (This option does not consider the fate of coastal canneries left behind.)

4. Selling dolphin-associated tuna through other outlets.

5. Perhaps sale of U.S. vessels to nations willing to continue dolphin fishing in the ETP.

The recent decline in the number of U.S. vessels that fish on dolphins in the ETP might be an early indication of the economic consequences of the canneries' decision.

Captain Performance Standards

The 1988 amendments to the MMPA mandate that the Secretary of Commerce develop and implement "a system of performance standards to maintain the diligence and proficiency of certificate holders" among the U.S. captains, the implication being that a similar system of performance standards would be expected of foreign fleets as well. The implementation of such an international system is now being developed through IATTC. NMFS also is developing a system whereby captains in the U.S. fleet would be penalized when their performance was below a given standard.

Training and Evaluation of Fishing Captains

Initial Training Most captains of tuna vessels learn their craft from another captain aboard a boat and serve as navigators for several years before being put in charge of a vessel. If they serve under a captain with a record of low dolphin mortality, they are likely to be well trained and motivated in this regard, although the degree of motivation also tends to depend on the individual.

In addition, some governments require that captains undergo more formal and specialized instruction on fishing laws and regulations, as well as in the use of fishing gear and techniques required for fishing for tuna in association with dolphins. Unless they comply with this requirement, captains are not issued the license that permits them to fish on dolphins. NMFS conducts such workshops for captains of U.S. vessels, with the following agenda:

- Introduction.
- Training requirements.
- Responsibilities of the Tuna-Dolphin Management Branch.
- The MMPA.
- Certificates of inclusion.
- International aspects of the tuna-dolphin problem.
- The observer program.
- Reporting requirements.
- Mortality reduction technology.
- Suggestions from the participants for improving dolphin safety.
- Marine mammal identification and geographical distribution.
- Quotas.
- Dolphin safety gear requirements.
- General procedural requirements.
- Dolphin safety and release procedures.
- Observers.

IATTC organizes workshops and seminars for captains, crews, and managers of tuna vessels in the international fleet. These workshops typically include:

- Activities of IATTC's international tuna-dolphin program.
- Historical review of the purse-seine fishery in the ETP.
- Historical review of incidental dolphin mortality.
- Fleet sampling and mortality estimation.
- International controversy over the tuna-dolphin program.
- Factors affecting dolphin mortality; responsibilities of captains and companies.
- Laws and regulations.
- Comparison of mortality rates, such as those between captains and between fleets.
- IATTC gear program and services; recent gear innovations.
- Review of dolphin safety gear.
- Meetings with individual captains to discuss their performances.

Measuring the Captains' Performance Some of the factors affecting dolphin mortality, such as gear malfunctions or subsurface currents, are beyond the control of captain and crew. However, the crew's skill and motivation play a major role in determining a vessel's dolphin mortality record. The best captains train their crews in the procedures for reducing dolphin mortality and instill in them the desire to do their best in every set to avoid killing dolphins. Even the best captains occasionally experience a set with high mortality, but in the long run their records show the difference (see Figure 6-3). From the point of view of management, captains who are responsible for the most mortality must be identified and measures must be taken, which may range from requiring additional training to revoking their fishing licenses.

To assess a captain's performance fairly, the effects of biases arising from fishing in different areas or on different stocks should be removed. This approach is used by IATTC. Alternatively, a simpler measure of performance may be used with the assumption that the effects of these factors will even themselves out over a long period of time for all captains. In 1989, NMFS proposed a system of the latter type, in which the average mortality per set from one trip would be compared with the average value for all captains. If the figure for a trip exceeded the average by 50% or more, certain actions would be taken. This system is being revised; new guidelines were not available for this report.

Remedial Actions Procedures have been developed for dealing with captains with records of high dolphin mortality. When such a captain has been identified, an expert-captain panel is convened. A group of experienced captains from a national fleet, both active and retired and with low dolphin mortality records, meets with the captain in question, reviews data from his trips, and discusses problems and solutions. This approach has proved to be effective, but it has some drawbacks. Small fleets may not have enough captains in port at any one time to convene such a panel, and new fleets may not have enough captains, active or retired, with sufficient experience to form an effective panel. The rapid expansion of foreign fleets in the 1980s created a need for vessel captains, and many were promoted from the ranks or brought in from fisheries in the Atlantic and Indian oceans. This development resulted in a large number of relatively inexperienced captains who were unfamiliar with fishing on dolphins and who are now in various stages of learning. Because of these problems, IATTC is providing the technical expertise for the assessment and correction of captain performance for the international fleet.

In the United States, the Porpoise Rescue Foundation tracks the performance of all U.S. fleet captains and intervenes when necessary. Other countries have their own national programs, usually coordinated through industry associations or specialized organizations, which monitor the performance of the captains of the national fleet. The procedures vary for dealing with captains with records of high dolphin mortality. The captain's record and the magnitude of the problems experienced are considered before action is taken. IATTC cooperates with all these national initiatives and provides data and technical support for their interpretation.

Both NMFS and IATTC offer services to the fleets for testing dolphin safety gear. These services involve a trial set to evaluate the condition and operation of the net and the other gear. The procedure is especially recommended when a captain is experiencing chronic problems that may be due to gear that is in poor condition or is used incorrectly. IATTC also offers set-by-set analyses of observed trips, detailing the use of dolphin safety equipment.

8

Recommendations

The tuna vessels fishing in the ETP and the governments under whose flags they fish have made significant progress in reducing dolphin mortality in the past two decades. In addition, three major tuna canneries recently announced that they would not buy or sell tuna caught in purse seines set around schools of dolphins, and other canneries may take similar steps. However, it was the clear consensus of this committee that the problem of reducing dolphin mortality from tuna fishing in the ETP is far from solved.

The effects of the voluntary ban by the canneries on actual dolphin mortality, for example, are uncertain. Enforcement of this ban may not be practical or even possible. For example, tuna fishermen may transship tuna caught on dolphins to foreign ports where it may be canned and sold to U.S. canneries as dolphin-free tuna. The ban may even increase mortality if it drives boats out of the more closely regulated nations' fisheries.

The committee's recommendations are divided into two parts. The first part recommends avenues for developing promising new techniques for reducing dolphin mortality in the existing purse-seine fishery on dolphins. The second part recommends research on and development of new methods of harvesting ETP yellowfin not in association with dolphins.

RECOMMENDATIONS CONCERNING THE ETP
TUNA-DOLPHIN FISHERY

These recommendations fall into four categories: (1) vessel-captain education, certification, incentives, and monitoring; (2) modifications of purse-seine gear and methods; (3) research on behavior of tuna and dolphins; and (4) development of new methods of harvesting yellowfin without encircling dolphins.

Vessel-Captain Education, Certification, Incentives, and Monitoring

The committee recommends that an international meeting be convened of representatives from government and industry from all countries engaged in the ETP purse-seine fishery. The purposes of this meeting would be the following:

1. Develop an educational certification and monitoring protocol for all captains in the international fleet, which would include:

- Establishment of program objectives.
- Training and evaluation of captain performance.
- Criteria for certification of captain.
- Monitoring through 100% observer coverage.

2. Initiate research on the development of incentives to improve captain performance, as discussed in Chapter 7.

The primary objective of this program would be to reduce dolphin mortality caused by the relatively small number of captains that account for a majority of the kill (see Figure 6-4). *The committee believes that improvement in captain performance is the single most important step that can be taken to reduce dolphin mortality in the ETP purse-seine fishery.* For example, if in 1989 the average kill per set for all captains of the international fleet had been the same as the average for those of the U.S. fleet, the total dolphin mortality would have been reduced by 60% (see Figure 6-3). This prediction is being fulfilled: In 1991 to date, the average mortality per set for the whole international fleet matched the 1989 U.S. rate, and the reduction in the total mortality between 1989 and the projected level for 1991 is 75%. If in 1989 all captains had operated with the rates of the best five captains of the international fleet, the total mortality would have been reduced by 87%. This reduction could occur without making any improvement in the basic technology or auxiliary equipment of purse seining.

Modifications of Purse-Seine Gear and Methods

The committee recommends that two approaches, short term and long term, be undertaken in gear and methods research and development.

First, a number of small modifications of current methods (see Chapter 7) could be built and tested immediately on commercial fishing trips. The most promising of these are the current profiler, jet boat, double corkline, pear-shaped snap rings, and polyester net. *The committee emphasizes that it is of paramount importance that an international program be developed to systematically deploy, test, and evaluate these modifications of current methods.*

Second, a number of major modifications of current methods need to be researched and developed on a long-term basis. Many of these are described in Chapter 7 and include such modifications as inflatable sections or partitions in the net, lifting surfaces, modified purse cable, new netting materials, and modified net designs. *The committee is recommending a long-term approach toward eliminating major causes of dolphin mortality in the purse-seine process—canopies, roll-downs, and collapses in the backdown channel—that couples an understanding of tuna and dolphin behavior with sound engineering.*

Some of the minor modifications to purse seines, as discussed in Chapter 7, would offer an incremental reduction in dolphin mortality and little risk or expense to the commercial enterprise. These improvements are recommended because they look promising based on previous experimentation. These modifications may result in significant reductions in dolphin mortality, but they should not be expected to achieve a kill rate approaching zero.

On the other hand, some of the major modifications discussed may hold the potential to reduce dolphin mortality significantly, but none of them is developed sufficiently to offer a clear advantage without some associated risk. This risk can come in several forms: financial risk to the operator, continued risk to dolphins, or risk to the tuna stocks. Examples of each type of risk are the cost of a new net of unproven advantage, the adoption of a dolphin-sparing gear having unanticipated dangers for the dolphins, and the adoption of a technique that results in exploiting immature tuna not associated with dolphins and thus reducing stocks.

Research is needed for a better understanding of the behavior of both dolphins and tuna and the bond between them. Details, costs, and potential benefits of many of the concepts delineated above cannot be judged at present.

Previous research programs addressing the tuna-dolphin problem have been constrained in three significant ways: (1) inadequate support of gear development and research on behavior; (2) the use of vessels in which the production of knowledge had to be balanced against the production of tuna; and (3) concentration on perfecting the conventional purse seine and the backdown process instead of developing other technologies. From a global perspective, the only way to reduce dolphin mortality is either to develop gear and techniques that will safely and efficiently harvest the tuna found beneath dolphins or to stop setting nets on dolphins. The current approach to the problem—monitoring the dolphin mortality and prodding the backdown process toward perfection—will almost certainly not achieve zero mortality.

Fishermen need both incentives and options to make further progress in reducing dolphin mortality. The individual fisherman alone cannot be expected to develop the options that offer significant improvements if they represent major changes to the present gear because fishermen who adopt gear that is likely to reduce dolphin mortality might incur economic losses. A program of research should be established to develop options and to demonstrate them to the industry. Such a program should include two facets:

• An experimental research program of innovative gear that would investigate performance and techniques. This program would have access to a modern commercial purse seiner as a dedicated vessel that would not be constrained by the normal pressures of tuna productivity. Because the capture of animals by fishing gear involves interactions between the animals and the gear, the research must include a program of behavioral studies focused on the reactions of both tuna and dolphins to fishing gear and other stimuli. Research techniques would include underwater video observation and acoustic sensing and tracking. Most of the research effort would be at sea.

• The information gained from the research above would then be used to develop rational purse-seine modifications and alternative harvesting methods based on the engineering requirements of the fishery. This focus would use analysis and modeling to develop, refine, and evaluate each concept to ensure a reasonable chance of success before it was attempted as a commercial prototype. After the analysis and modeling, promising prototypes would be field-tested.

The tools of this research would include remote-operated vehicles (ROVs), scanning sonar, electronic gear monitoring sensors, hydrodynamic test tanks, and computers, in addition to the dedicated vessel. To accomplish the research, experts from a variety of disciplines would be used. In addition to people with a historical involvement in the tuna fishery, experts from other fisheries and from relevant engineering and scientific disciplines would be involved. The goal of the research would be to develop techniques that can exploit yellowfin tuna in the ETP without significant dolphin mortality. The program would be of finite duration, and long-term monitoring activities would be performed by existing agencies. The program would cost at least several million dollars.

Some specific aspects of the interaction of tuna, dolphins, and fishing gear that are in particular need of study are discussed below.

Tuna and Dolphin Behavior in and Around the Seine

As was indicated in Chapter 5, very little is known about tuna and dolphin behavior, particularly when they are inside the net and when they escape from the net. Either scanning sonar or ROVs could be used in conjunction with a chartered purse seiner to study this behavior. Several scientists have proposed using ROVs to allow videotaping and other sensing during tuna seining. These mobile data-gathering platforms can be designed to operate in the net, even into the backdown period. Their use throughout the entire course of the set should allow the gathering of many visual and acoustical data showing the relations of tuna and dolphins to each other and to the net. It should also be possible to gather data during the seine set by deploying the ROV from a small associated vessel and then setting the net around both vessel and ROV.

ROV technology could help answer some of the following questions:

- Is the tuna-dolphin bond broken during seining? If so, when and why?
- How deep in the net do tuna go?
- Is there any useful behavioral difference between the two partners in the bond that would allow differential release of dolphins and retention of tuna?
- When might this separation occur in a set?
- What is the underwater effect of outside stimuli such as net skiffs, jet skis, and speedboats on dolphins and fish?
- What are the shape, behavior, and disposition of dolphins and tuna schools in a net at the various stages of its closing?
- What is the relation of tuna and dolphin to the net throughout a set?
- Can more be learned about entanglement, avoidance of canopies, and other problems of backdown?
- Can changes in dolphin behavior at sunset, compared with daylight, be documented?

In summary, thorough videotaping of events throughout a set should provide a great deal of new information about behavior during seining.

The Effects of Chasing Dolphin Herds

No specific information is available concerning the effects of the chase on the biology of dolphins. The chase is likely to result in stress. Some herds have developed strategies to avoid capture; others seem to have habituated to encirclement and seem to have developed behavioral patterns that reduce their risks once in the net. Further studies on physiological and behavioral impacts of the chase are obviously needed.

Sunset Set Tests

The increased mortality of seined dolphins taken near dusk makes it seem likely that visual problems are involved. Behavior of netted dolphins as seen from ROVs probably would be instructive. Measurements of the progress of light intensity in the net as dusk arrives would also be useful, as would testing of net materials that could improve the visibility of the net.

Bubble Tests

If, as is suspected, bubble trails from the seiner's propeller are crucial to the present success of encircling dolphins, it may be that bubbles made by generators could be used to maneuver animals in the net and hence to help release them. Such methods could reduce the duration of sets and hence the stress upon netted animals and aid in pre-backdown release.

The Potential for Maneuvering Tuna and Dolphins Inside the Seine and Dolphin Release-Gate Tests

If current models of dolphin-herd function are correct, as discussed in Chapter 7, then it should be possible to predict the size of openings that will be required to exploit normal escape behavior of dolphins to produce quick and complete release of dolphins from tuna seines.

Three needs must be met if this concept can be used to release dolphins during tuna seining. First, the mechanical and operational means of producing such a gate at sea without loss of tuna is essential. Second, the predicted dimensions of such an opening must be refined in actual tests. Third, proposed methods of maneuvering dolphin herds in the open net toward such a gate must be tested and refined.

Research on Behavior of Tuna and Dolphins

Oceanographic Correlates

If the association of tuna with dolphins is an extension of the behavior that leads them to associate with flotsam, as some researchers believe, comparative data are needed on environmental conditions where dolphins and tuna associate and where they do not associate. All observers on seiners should be trained and equipped to collect basic environmental measurements, including water color and clarity, thermal profiles, and weather, in all sets, whether on dolphins, flotsam, or school fish. Good samples of the same information should be collected from places without fish.

Simultaneous Tracking of Dolphins and Associated Yellowfin

The most direct way to determine the durability of the tuna-dolphin bond is to tag and follow individuals of both species that are caught together. Carey and Olson (1981) tagged yellowfin tuna with ultrasonic transmitters and successfully followed individuals. They uncovered enough information about tuna behavior that was not suspected to show that the effort needs to be extended and coupled with simultaneous radio tracking of tuna and dolphins. Recent advances in telemetry allow measurement of enough factors, such as body temperature, swimming speed, depth, and duration of dives, to construct a good profile of the environmental correlates of the bond.

Tracking Flotsam-Associated Yellowfin Tuna

Placing sonic tags on tuna caught near flotsam or FADs would provide valuable data on that association as well as data for comparison with those obtained from dolphin-associated tuna. Comparative data from large yellowfin tuna associated with flotsam are especially important.

The committee believes that IATTC's flotsam information program—aimed at finding out why tuna associate with logs and with dolphins (see Appendix 2)—should be intensified. All IATTC observers document the physical and biological characteristics of the flotsam they find. In 1987, they attached tags to the objects and sent out notices to fishermen requesting documentation of the times and places where fishermen saw a tagged log, its condition, and information about any fish associated with it (see Appendix 3). The objective was to understand drift patterns and learn about longevity and relative attractiveness of various kinds of flotsam. Recoveries were very few, but IATTC hopes to repeat the experiment, using satellite transmitters to track the logs. All observers on seiners should collect these data, as should oceanographic research vessels. Body-length data for the fish caught in different situations and at different locations also are important to collect.

Studies Using Fish-Aggregating Devices

Although tuna aggregation around floating objects differs in important ways from tuna aggregation with dolphins, the tendencies that underlie tuna aggregation around logs and dolphins probably have much in common. Studies using instrumented FAD platforms are likely to provide valuable information. A FAD can be easily designed as a data-gathering platform, and instrumented with lights and 24-hour underwater video and other monitoring devices (such as acoustic sampling). Sets made around instrumented FADs may reveal further information about the reactions of animals to seining. In addition, the use of structures much larger than current FADs might be worth consideration.

Satellite Monitoring of Radio-Tagged Dolphins

It is possible to track radio-tagged dolphins of various species for several days. Such work should illuminate the nature of the association between dolphin species—for example, spotted and spinner dolphins—taken in the same net. The degree of permanence of the herds that have experienced netting and the timing of their different feeding strategies should be defined. Disruption of schools caused by seining may have serious social consequences on dolphins, which could be monitored. Such data may begin to define a number of unknown features of dolphin herds and their interrelations, such as range of species. Long-term tracking should give information on the coherence of dolphin herds (i.e., are they persistent or are they frequently re-found with different individuals?) and data on the frequency of seining on specific dolphin herds, as well as their behavior during approach of vessels and after release. Such questions as how rapidly netted dolphins swim before they reaggregate into feeding herds and how far they travel can be considered. Short-term tracking may allow physiological assessments of the stress of seining on dolphins. Studies of the composition of netted herds also may help answer questions about the social consequences of seining on dolphins.

RECOMMENDATIONS FOR HARVESTING TUNA NOT IN ASSOCIATION WITH DOLPHINS

The committee recommends two avenues of research. *The first recommendation involves research into the night behavior of tuna and dolphins and the second involves research into new methods of purse seining large yellowfin not in association with dolphins in the ETP.* If purse seining for tuna in association with dolphins is discontinued, the committee recommends that the impact on tuna populations be carefully assessed and further notes that additional restrictions may be necessary to maintain the productivity and viability of tuna stocks in the ETP. Differences exist in the size-specific fishing mortality rates imposed by dolphin sets and non-dolphin sets, the result being that sets of the latter type produce lower yield per recruit than the former. Although a relationship between stock size and recruitment has not been detected for the range of observed population levels, increased pressure on young tuna eventually may reduce adult stock size to a level at which recruitment may be reduced, which would further decrease overall production.

Night Behavior of Tuna and Dolphins

If large yellowfin do not associate with dolphins at night, purse seining or trawling could be done at night, which would reduce dolphin mortality significantly. To study night behavior, tuna and dolphins that are associated during the day need to be tracked at night, the tuna with acoustic tags and the dolphins with radio tags.

New Methods of Purse Seining Large Yellowfin
Without Encircling Dolphins

Even if harvest of yellowfin tuna in association with dolphins continues, the committee recommends that a major research effort be undertaken to explore new methods of harvesting yellowfin in the ETP. Three promising avenues of research identified by the committee are as follows:

• Can anything about FADs be changed—e.g., size, depth, shape, structure, composition—to make them attract larger yellowfin and change the mixture of species they attract? In the short term, existing FADs could be deployed and monitored by observer-manned commercial fishing vessels. In the long term, a chartered purse seiner could be used to investigate the performance of such new technologies as submerged FADs, which may have a greater potential than surface FADs for attracting and holding commercially harvestable schools of large yellowfin tuna. This research would be conducted in conjunction with that described in Chapter 5 on the behavior of tuna and dolphins.

• The use of satellite oceanographic techniques to locate aggregations of tuna not associated with dolphins in the ETP. Information from satellites on surface oceanography would be analyzed with catch information from tuna-vessel log books to determine whether large yellowfin aggregate in harvestable numbers without dolphins under certain environmental conditions in the ETP.

• The use of alternative techniques of locating tuna that do not depend on the sighting of dolphins. Emerging technologies such as light-induced detecting and ranging (LIDAR) and synthetic aperture radar (SAR) hold promise because of their ability to detect the presence of tuna as described in Chapter 7; sonar, despite its drawbacks, might have promise as well.

In addition to the above three items, the committee members had diverse opinions about the promise of research on the potential of non-purse-seine techniques that do not involve the encircling of dolphins. Two methods were discussed. The first would be the development of high-intensity economically viable techniques of longlining such as the use of a fleet with catcher boats and a mother vessel (which works in other fisheries) or the use of helicopters to place baited hooks in the path of tuna schools (which is more remote a possibility). The second would be development of midwater pair-trawl techniques designed to catch tuna by towing well beneath the associated dolphins.

SUMMARY

In summary, the committee recommends that two major international efforts be undertaken to reduce the mortality of dolphins in the ETP tuna fishery. Once again, we emphasize the word "international," since most of the ETP dolphin mortality occurs in the non-U.S. fishery. The first recommendation is to develop an educational, monitoring, and incentives program for tuna vessel captains aimed at reducing the dolphin mortality from the relatively small number of captains that accounts for a large proportion of the kill. The committee considers this to be the single most important short-term step that can be taken to reduce dolphin mortality in the existing ETP purse-seine fishery. The second recommendation is to develop a major international program for gear and behavior research aimed at reducing dolphin mortality through the following:

• Systematic deployment of small, currently available modifications of present-day purse-seine methods that show promise (e.g., current profiler, jet boat, double corkline, pear-shaped snap rings, and polyester net).

• Development of major new modifications in purse-seine technology through a three-pronged program of engineering, modeling, and full-scale testing using a chartered purse seiner.

• Design and implementation of a major research program on the behavior of tuna and dolphins in the ETP. This program would be conducted from a chartered purse seiner and would investigate such phenomena as (1) tuna and dolphin behavior when they are inside the purse-seine net and when they are escaping from the net; (2) night behavior of tuna and dolphins; (3) the effects of chasing dolphin herds; (4) the effects of sets in dim light; and (5) techniques for maneuvering and separating tuna and dolphins inside the seine.

• Design and implementation of a major research program on new methods of harvesting yellowfin in the ETP without encircling dolphins. This program would be conducted from a chartered purse seiner and would investigate such concepts as (1) the use of surface and subsurface FADs to aggregate large yellowfin; (2) the use of satellite oceanographic techniques to locate aggregations of large yellowfin not in association with dolphins in the ETP; and (3) the use of LIDAR and SAR, and possibly sonar, to detect tuna.

The committee recommends that both programs be initiated immediately. The vessel captain educational, monitoring, and incentives program should be created through international meetings of government and industry representatives of all nations currently purse seining in the ETP. To implement and evaluate this effort, mortality reduction targets must be established. The gear and behavior research program should be developed through a collaborative international effort and should be designed to run for 5-7 years.

References

Allen, R.L., and M.D. Goldsmith. 1981. Dolphin mortality in the eastern tropical Pacific incidental to purse seining for yellowfin tunas, 1979. Rep. Int. Whal. Comm. 31:539–540.

Allen, R.L., and M.D. Goldsmith. 1982. Dolphin mortality in the eastern tropical Pacific incidental to purse seining for yellowfin tuna, 1980. Rep. Int. Whal. Comm. 32:419–421.

Alverson, F.G. 1963. The food of the yellowfin and skipjack in the eastern tropical Pacific Ocean. Inter-Am. Trop. Tuna Comm., Bull. 7:293–396.

Anganuzzi, A.A., and S.T. Buckland. 1989. Reducing bias in trends in dolphin relative abundance, estimated from tuna vessel data. Rep. Int. Whal. Comm. 39:323–334.

Anganuzzi, A.A., S.T. Buckland, and K.L. Cattanach. In press, a. Relative abundance of dolphins associated with tuna in the eastern tropical Pacific, estimated from tuna vessel sightings data for 1988 and 1989. IWC Doc. SC/42/CM36. Rep. Int. Whal Comm.

Anganuzzi, A.A., K.L. Cattanach, and S.T. Buckland. In press, b. Relative abundance of dolphins associated with tuna in the eastern tropical Pacific, estimated from preliminary tuna vessel sightings data for 1990. IWC Doc. SC/43/SM14. Rep. Int. Whal. Comm.

Anon. 1990. Drift Nets Catch Three Dolphins per Cast. Fisheries Agency. The Japan Times Weekly, International Edition, March 5–11.

Aron, W. 1988. The commons revisited: Thoughts on marine mammal management. Coastal Manage. 16:99–110.

Au, D., and W. Perryman. 1977. Oceanography of Pelagic Dolphins. Proceedings of Second Conference on Biology of Marine Mammals, December 12–15, 1977, Naval Ocean Systems Center, San Diego, Calif.

Au, D., and W. Perryman. 1982. Movement and speed of dolphin schools responding to an approaching ship. Fish. Bull. 80:371–379.

Au, D., and R.L. Pitman. 1988. Seabird relations with tropical tunas and dolphins. Pp. 174–212 in Seabirds and Other Marine Vertebrates, J. Burger, ed. New York: Columbia University Press.

Au, D., W. Perryman, and W.F. Perrin. 1979. Dolphin Distribution and the Relationship to Environmental Features in the Eastern Tropics. Southwest Fisheries Center Administrative Report No. LJ-79-43. National Marine Fisheries Service, Southwest Fisheries Center, La Jolla, Calif. 59 pp.

Bratten, D.A., and R. Guillen. 1981. Cruise Report on Local Testing Aboard the M.V. SANTA ELENA. Inter-American Tropical Tuna Commission, La Jolla, Calif.

Bratten, D.A., W. Ikehara, K. Pryor, P. Vergne, and J. DeBeer. 1979. The Tuna/Porpoise Problem: Dedicated Vessel Research Program. Summary of Research Results from the Second Leg of the Third Cruise of the Dedicated Vessel, July 20–August 18, 1978. Southwest Fisheries Center Administrative Report No. LJ-79–13. National Marine Fisheries Service, Southwest Fisheries Center, La Jolla, Calif. 52 pp.

Buckland, S.T., and A.A. Anganuzzi. 1988. Estimated trends in abundance of dolphins associated with tuna in the eastern tropical Pacific. Rep. Int. Whal. Comm. 38:411–437.

Buckland, S.T., K.L. Cattanach, and A.A. Anganuzzi. In press. Estimating trends in abundance of dolphins associated with tuna in the eastern tropical Pacific Ocean, using sightings data collected on commercial tuna vessels. Fish. Bull.

Burnham, K.P., D.R. Anderson, and J.L. Laake. 1980. Estimation of density from line transect sampling of biological populations. Wildl. Monogr. 72. 202 pp.

Caldwell, D.K., and M.C. Caldwell. 1971. Porpoise Fisheries in the Southern Caribbean: Present Utilization and Future Potentials. Rosenstiel School Marine and Atmospheric Science. Pp. 195–205 in Proceedings of 23rd Annual Session, Gulf and Caribbean Fishery Institute, June 1971.

Carey, F.G., and R.J. Olson. 1981. Sonic Tracking Experiments with Tunas. Paper presented at Symposium Definition of Tuna and Billfish Habitats and Effects of Environmental Variations on Apparent Abundance and Vulnerability to Fisheries, Tenerife, Spain. 18 pp.

Cayre, P., J.B. Amon Kothias, T. Diouf, and J-M. Stretta. 1988. Biologie des Thons. Ressources, Pche et Biologie des Thonids Tropicaux de l'Atlantique Centre-Est, Fonteneau, A., and J. Marcille, eds. FAO Document Technique sur les Pches 292.

Cicin-Sain, B., M. Orbach, S. Sellers, and E. Manzanilla. 1986. Conflictual Interdependence: United States-Mexico Relations on Fishery Resources. Nat. Resour. J. 26:769–792.

Clay, C.S., and H. Medwin. 1972. Acoustical Oceanography: Principles and Applications. New York: John Wiley & Sons.

Coan, A.L., and G.T. Sakagawa. 1982. An examination of single set data for the U.S. tropical tuna purse seine fleet. Pp. 83–94 in ICCAT, Collective Volume of Scientific Papers, Vol. 18, SCRS/82/53.

Coe, J.M., D.B. Holts, and R.W. Butler. 1984. The "tuna-porpoise" problem: National Marine Fisheries Service dolphin mortality reduction research, 1970–81. Mar. Fish. Rev. 46:18–33.

Cole, J.S. 1980. Synopsis of biological data on the yellowfin tuna, *Thunnus albacares*, (Bonnaterre, 1788), in the Pacific Ocean. Pp. 71–150 in Synopses of Biological Data on Eight Species of Scombroids, W.H. Bayliff, ed. Inter-Am. Trop. Tuna Comm. Spec. Rep. No. 2. 530 pp.

Connor, R.C. 1987. Aggressive Herding of Females by Coalitions of Male Bottlenose Dolphins (Tursiops sp.). 7th Biennial Conf. Biol. Marine Mammals, Miami, Fla. (Abstract)

Dailey, M.D., and W.F. Perrin. 1973. Helminth parasites of porpoises of the *Stenella* in the tropical eastern Pacific, with descriptions of two new species: *Mastigonema stenellae* gen et sp. N. (Nematoda; Spiruroidea) and *Zalophotrema pacificum* sp. N. (Trematoda; Digenea). Fish. Bull. 71:455–471.

DeBeer, J. 1980. Cooperative Dedicated Vessel Research Program on the Tuna-porpoise Problem: Overview and Final Report, Contract No. MM8AC006. Marine Mammal Commission, Washington, D.C.

DeFran, R.H., and K.W. Pryor. 1980. The behavior and training of cetaceans in captivity. Pp. 319–362 in Cetacean Behavior, Mechanisms and Functions, L.M. Herman, ed. New York: John Wiley & Sons.

DeMaster, D.P., ed. 1989. Workshop on Alternative Methods to Purse-seining for Yellowfin Tunas in the Eastern Tropical Pacific, October 11–12, 1988. Southwest Fisheries Center Administrative Report No. LJ-89–96. National Marine Fisheries Service, Southwest Fisheries Center, La Jolla, Calif. 21 pp.

Dolan, H. 1980. Public Hearings on Proposed Regulations Governing the Take of Marine Mammals Incidental to Commercial Tuna Fishing Operations. MMPAH 1980–1. March 1980.

Dolar, M.L.L. 1990. Interaction between Cetaceans and Fisheries in the Visayao, Philippines, a Preliminary Study. IWC Workshop on Mortality of Cetaceans in Passive Fishing Nets and Traps, La Jolla, Calif., October 22–25, 1990. 12 pp.

Edwards, E., and C. Glick. In press. Trend and Power in Dolphin Abundance Estimates Derived from Tuna Vessel Observer Data: Ten Year Intervals. NMFS SWFC Administrative Report.

Efron, B. 1982. The Jackknife, the Bootstrap and other Resampling Plans. SIAM, Monogr. No. 38, CBMS-NSF.

FAO (Food and Agricultural Organization). 1984. Yearbook of Fishery Statistics. Food and Agricultural Organization of the United Nations. Rome, Italy.

FAO (Food and Agricultural Organization). 1989. Yearbook of Fishery Statistics. Food and Agricultural Organization of the United Nations. Rome, Italy.

Fitch, J.E., and R.L. Brownell, Jr. 1968. Fish otoliths in cetacean stomachs and their importance in interpreting feeding habits. J. Fish. Res. Board Can. 25:2561-2574.

Fox, W.W., and W. Lenarz. 1975. Progress of Research on Porpoise Mortality Incidental to Tuna Purse Seine Fishing for Fiscal Year 1975. Southwest Fisheries Center Administrative Report No. LJ-75–68. National Marine Fisheries Service, Southwest Fisheries Center, La Jolla, Calif. 73 pp.

Francis, R.C. 1986. Two fisheries biology problems in west coast groundfish management. North Am. J. Fish. Manage. 6:453–462.

Gerrodette, T., and P. Wade. In press, a. Monitoring trends in dolphin abundance in the eastern tropical Pacific: Analysis of 1989 data. IWC Doc. SC/42/SM00. Rep. Int. Whal. Comm.

Gerrodette, T., and P. Wade. In press, b. Monitoring trends in dolphin abundance in the eastern tropical Pacific: Analysis of five years of data. IWC Doc. SC/43/SM13. Rep. Int. Whal. Comm.

Glotov, V.P. 1962. Investigation of the scattering of sound by bubbles generated by an artificial wind in seawater and the statistical distribution of bubble sizes. Sov. Phys.-Acoust. 7:341–345. (Russian)

Goudey, C.A. 1987. New rig gives added height to bottom trawl. Commercial Fisheries News, February.

Greenblatt, P.R. 1979. Associations of tuna with flotsam in the eastern tropical Pacific. Fish. Bull. 77:147–155.

Guillen, R., and D.A. Bratten. 1981. Anchored Raft Experiment to Aggregate Tunas in the Eastern Tropical Pacific Ocean. Internal Report No. 14, Inter-American Tropical Tuna Commission, La Jolla, Calif. 13 pp.

Hall, M.A., and S.D. Boyer. 1986. Incidental mortality of dolphins in the eastern tropical Pacific tuna fishery: Description of a new method and estimation of 1984 mortality. Rep. Int. Whal. Comm. 36:375-381.

Hall, M.A., and S.D. Boyer. 1987. Incidental mortality of dolphins in the eastern tropical Pacific tuna fishery in 1985. Rep. Int. Whal. Comm. 37:361–362.

Hall, M.A., and S.D. Boyer. 1988. Incidental mortality of dolphins in the eastern tropical Pacific tuna fishery in 1986. Rep. Int. Whal. Comm. 38:439–441.

Hall, M.A., and S.D. Boyer. 1989. Incidental mortality of dolphins in the eastern tropical Pacific tuna fishery in 1987. Rep. Int. Whal. Comm. 39:321–322.

Hall, M.A., and S.D. Boyer. 1990. Incidental mortality of dolphins in the tuna purse-seine fishery in the eastern Pacific Ocean during 1988. Rep. Int. Whal. Comm. 39:321–322.

Hall, M.A., and S.D. Boyer. In press, a. Incidental mortality of dolphins in the tuna purse-seine fishery in the eastern Pacific Ocean during 1989. Rep. Int. Whal. Comm. 41.

Hall, M.A., and S.D. Boyer. In press, b. Estimates of incidental mortality of dolphins in the purse-seine fishery for tunas in the eastern Pacific Ocean in 1990. Rep. Int. Whal. Comm. 42.

Hall, M.A., F.E. Mann, and M.D. Scott. In press. Behavioral Adaptations of Dolphins Affected by the Tuna Purse-Seine Fishery. Inter-American Tropical Tuna Commission, Scripps Institution of Oceanography, La Jolla, Calif.

Hallier, J.P. 1985. Purse seining on debris-associated schools in the western Indian Ocean. Pp. 150–164 in Expert Consultation on Stock Assessment of Tunas in the Indian Ocean, Columbo, Sri-Lanka, 28 November–2 December 1985. FAO Document No. TWS/85/30. Food and Agricultural Organization of the United Nations, Rome, Italy.

Hammond, P.S. 1984. Dolphin mortality incidental to purse-seining for tunas in the eastern tropical Pacific Ocean, 1982. Rep. Int. Whal. Comm. 34:539–541.

Hammond, P.S., and M.A. Hall. 1985. Dolphin mortality incidental to purse-seining for tunas in the eastern tropical Pacific inflicted by the U.S. fleet in 1983 and non-U.S. fleet in 1979-1983. Rep. Int. Whal. Comm. 35:431–433.

Hammond, P.S., and J.L. Laake. 1983. Trends in estimates of abundance of dolphins (*Stenella* spp. and *Delphinus delphis*) involved in the purse-seine fishery for tunas in the eastern Pacific Ocean, 1977–1981. Rep. Int. Whal. Comm. 34:565–583.

Hammond, P.S., and K.T. Tsai. 1983. Dolphin mortality incidental to purse-seining for tunas in the eastern Pacific Ocean, 1979–1981. Rep. Int. Whal. Comm. 33:589–597.

Hasler, A.D. 1966. Homing of Salmon; Underwater Guideposts. Madison: University of Wisconsin Press. 155 pp.

Hofman, R.J. 1979. A Workshop to Identify New Research That Might Contribute to the Solution of the Tuna-Porpoise Problem. Marine Mammal Commission, Washington, D.C.

Hofman, R.J., ed. 1981. Identification and Assessment of Possible Alternative Methods for Catching Yellowfin Tuna. Sponsored by the Marine Mammal Commission in cooperation with the Inter-American Tropical Tuna Commission and the National Marine Fisheries Service. Report No. PB83–138933. Springfield, Va.: National Technical Information Service.

Hoge, F.E., C.W. Wright, W.B. Krabill, R.B. Buntzen, G.D. Gilbert, R.N. Swift, J.K. Yungel, and R.E. Berry. 1988. Airborne lidar detection of subsurface oceanic scattering layers. Appl. Opt. 27:3969–3977.

Holbrook, J.R. 1980. Dolphin Mortality Related to the Yellowfin Tuna Purse Seine Fishery in the Eastern Tropical Pacific: An Annotated Bibliography. Tech. Bull. 2, Porpoise Rescue Foundation, San Diego, Calif. 131 pp.

Holt, R.S., and J.E. Powers. 1982. Abundance Estimation of Dolphin Stocks Involved in the Eastern Tropical Pacific Yellowfin Tuna Fishery Determined from Aerial and Ship Surveys to 1979. NOAA Tech. Memo. 23. National Marine Fisheries Service, Southwest Fisheries Center, La Jolla, Calif. 95 pp.

IATTC (Inter-American Tropical Tuna Commission). 1987. Annual Report: 1986. Inter-American Tropical Tuna Commission, La Jolla, Calif.

IATTC (Inter-American Tropical Tuna Commission). 1989a. The Fishery for Tunas in the Eastern Pacific Ocean. Tuna-Dolphin Workshop, San Jose, Costa Rica, March 14–16, 1989, Working Document No. 1. La Jolla, Calif.: Scripps Institution of Oceanography.

IATTC (Inter-American Tropical Tuna Commission). 1989b. Annual Report: 1988. Inter-American Tropical Tuna Commission, La Jolla, Calif.

IATTC (Inter-American Tropical Tuna Commission). 1989c. Incidental Mortality of Dolphins in the Eastern Tropical Pacific Tuna Fishery, 1979–1988. A Decade of the Inter-American Tropical Tuna Commission's Scientific Technician Program. Tuna-Dolphin Workshop, San Jose, Costa Rica, March 14–16, 1989, Working Document No. 2. La Jolla, Calif.: Scripps Institution of Oceanography.

IATTC (Inter-American Tropical Tuna Commission). 1989d. Assessment Studies of Yellowfin Tuna in the Eastern Pacific Ocean. Background Paper 2, 46th Meeting of the IATTC, May 10–12, 1989, La Jolla, Calif. 42 pp.

IATTC (Inter-American Tropical Tuna Commission). 1989e. The Gear Program. Tuna-Dolphin

Workshop, San Jose, Costa Rica, March 14–16, 1989, Working Document No. 4. La Jolla, Calif.: Scripps Institution of Oceanography.

IATTC (Inter-American Tropical Tuna Commission). 1990a. Assessment Studies of Yellowfin Tuna in the Eastern Pacific Ocean. Background Paper 2, 47th Meeting of the IATTC, June 26–28, 1990, La Jolla, Calif.

IATTC (Inter-American Tropical Tuna Commission). 1990b. Quarterly Report: April–June, 1990. Inter-American Tropical Tuna Commission, La Jolla, Calif.

IATTC (Inter-American Tropical Tuna Commission). 1991a. Annual Report: 1989. Inter-American Tropical Tuna Commission, La Jolla, Calif.

IATTC (Inter-American Tropical Tuna Commission). 1991b. Assessment Studies of Yellowfin Tuna in the Eastern Pacific Ocean. Background Paper 2, 49th Meeting of the IATTC, June 18–20, 1991, La Jolla, Calif.

Iverson, R.T.B. 1987. U.S. Tuna Processors. The Development of the Tuna Industry in the Pacific Islands Region: An Analysis of Options, D.J. Doulman, ed. Hawaii: East-West Center.

IWC (International Whaling Commission). In press. Report of the 43rd International Whaling Commission Meeting, Reykjavik, Iceland, May 1991.

Joseph, J., and J. Greenough. 1979. International Management of Tuna, Porpoise and Billfish: Biological, Legal and Political Aspects. Seattle: University of Washington Press.

King, D.M., and H.A. Bateman. 1985. The Economic Impact of Recent Changes in the U.S. Tuna Industry. Sea Grant Publication No. P-T-47. University of California, La Jolla.

Kirkland, B. 1990. Tuna Fishing Efficiency Scores Big Advance with Netting Technology. Sales Brochure, West Coast Netting, Rancho Cucamonga, Calif.

Lang, T.G. 1966. Hydrodynamic analysis of cetacean performance. Pp. 410-431 in Whales, Dolphins and Porpoises, K.S. Norris, ed. Berkeley: University of California Press.

Leatherwood, S., and D.K. Ljungblad. 1979. Nighttime swimming and diving behavior of a radio-tagged spotted dolphin, *Stenella attenuata*. Cetology 34:1–6.

Leatherwood, S., and R.R. Reeves. 1989. Marine Mammal Research and Conservation in Sri Lanka, 1985–1986. United Nations Environment Programme, Marine Mammal Technical Report Number 1, Nairobi, Kenya.

Levenez, J.J., A. Fonteneau, and R. Regalado. 1980. Resultats d'une Enquete sur L'importance des Dauphins dans la Pêcherie Thonèire FISM. Collective Volume of Scientific Papers, ICCAT/Recl. Doc. Sci. CICTA/Colecc. Doc. Cient. CICAX 9(l):176–179.

Living Marine Resources, Inc. 1982. Gulf Coast Tuna Resource Survey, April 1 to August 12, 1982, F/V Providence. Silva Fishing Company, Grant-in-Aid Award No. NA82-GA-D-005 15. 37 pp.

Lo, N.C.H., and T.D. Smith. 1986. Incidental mortality of dolphins in the eastern tropical Pacific, 1959-1972. Fish. Bull. 84:27–34.

Maigret, J. 1981. Introduction a l'etude des rapports entre cetaces et la pêche thonèire dans l'Atlantique tropical. Bulletin du Centre National de Recherches Océanographiques et des Pêches de Mouadhibou 10(1):89-101.

Maigret, J. 1990. Relationship Between Marine Mammals and the Fisheries on the Western African Coasts. Working Paper 5 of the IWC Conference on Mortality of Cetaceans in Passive Fishing Nets and Traps, October 1990, San Diego.

McFarland, W.N., and E.R. Loew. 1983. Wave produced changes in underwater light and their relations to vision. Environ. Biol. Fish. 8:173–184.

McFarland, W.N., and F.W. Munz. 1975. Part II: The photic environment of clear tropical seas during the day. Vision Res. 15:1063-1070.

McKnight, J.R. 1979. Commercial Fishing Net Assembly with a Porpoise Escape Zone. U.S. Patent 4,174,582. U.S. Patent Office, Washington, D.C.

McNeely, R.L. 1961. Purse seine revolution in tuna fishing. Pac. Fish. 59:27-58.

Mickey, M.R. 1959. Some finite population unbiased ratio and regression estimators. J. Am. Stat. Assoc. 54:594–612.

Mitchell, E. 1975. Porpoise, Dolphin and Small Whale Fisheries of the World: Status and Problems. IUCN Monograph No. 3. International Union for Conservation of Nature and Natural Resources, Morges, Switzerland. 129 pp.

MMC (Marine Mammal Commission). 1991. Annual Report of the Marine Mammal Commission: Calendar Year 1990. A Report to Congress. Washington, D.C.: Marine Mammal Commission.

Montaudouin, X., J.P. Hallier, and S. Hassani. 1990. Analyse des donnes collectes lors des embarquements bord des senneurs bass aux Seychelles (1986-1989). Seychelles Fishing Authority, June 1990.

NMFS (National Marine Fisheries Service). 1980a. Final Environmental Impact Statement. Proposed Regulations and Conditions for a General Permit Governing the Taking of Marine Mammals Associated with Tuna Purse Seine Operations, November 1980. National Marine Fisheries Service, Southwest Fisheries Center, La Jolla, Calif.

NMFS (National Marine Fisheries Service). 1980b. Fisheries of the United States: 1979. Current Fisheries Statistics No. 8000. Washington, D.C: National Marine Fisheries Service.

NMFS (National Marine Fisheries Service). 1986. A Review of Procedures to Develop Alternatives to Tuna Purse Seining on Porpoise. National Marine Fisheries Service, National Oceanic and Atmospheric Administration, U.S. Department of Commerce, Washington, D.C.

NMFS (National Marine Fisheries Service). 1987. Environmental Assessment on the Proposed Modification of the Marine Mammal Regulations Regarding the Importation of Yellowfin Tuna Caught with Purse Seines in the Eastern Tropical Pacific.

Norris, K.S. 1991. Dolphin Days: The Life and Times of the Spinner Dolphin. New York: Norton Press. 335 pp.

Norris, K.S., and T.P. Dohl. 1980a. The behavior of the Hawaiian spinner dolphin. Fish. Bull. 77:821–849.

Norris, K.S., and T.P. Dohl. 1980b. The structure and functions of cetacean schools. Pp. 211–261 in Cetacean Behavior: Mechanisms and Processes, L.M. Herman, ed. New York: John Wiley & Sons.

Norris, K.S., and C.R. Schilt. 1987. Cooperative societies in three-dimensional space: On the origins of aggregations, flocks, and schools, with special reference to dolphins and fish. Ethol. Sociobiol. 9:149-179.

Norris, K.S., W.E. Stuntz, and W. Rogers. 1978. The Behavior of Tuna and Porpoises in the Eastern Tropical Pacific Yellowfin Tuna Fishery: Preliminary Studies. Final Report for Marine Mammal Commission. Contract No. MM6AC022, Publication No. PB-283–970. Springfield, Va.: National Technical Information Service. 86 pp.

Norris, K.S., B. Wursig, R.S. Wells, M. Wursig, S.M. Brownlee, C. Johnson, and J. Solow. 1985. The behavior of the Hawaiian spinner dolphin *Stenella longirostris*. Southwest Fisheries Center Administrative Report No. LJ-85–06C. National Marine Fisheries Service, Southwest Fisheries Center, La Jolla, Calif.

Norris, K.S., B. Wursig, R.S. Wells, M. Wursig, S.M. Brownlee, C. Johnson, and J. Solow. In press. The Hawaiian Spinner Dolphin. Berkeley: University of California Press.

Northridge, S.P. 1984. World Review of Interactions between Marine Mammals and Fisheries. FAO Fish. Tech. Paper No. 251. FAO, Rome. 190 pp.

Northridge, S.P. 1991. An Updated World Review of Interactions Between Marine Mammals and Fisheries. FAO Fish. Tech. Paper No. 251, Suppl. 1. FAO, Rome. 58 pp.

Nye, R.D. 1990. Spectra Fiber in Commercial Fishing. Proceedings MTS 1990, Washington, D.C.

Orbach, M.K., and J.R. Maiolo. 1989. United States tuna policy: A critical assessment. Mar. Policy Rep. 1:307–332.

Pacific Tuna Development Foundation. 1977. Final Report. Tuna Purse-seine Charter to the Western Pacific, July–November, 1976. Pacific Tuna Development Foundation, Honolulu.

Pascual, J.N. 1961. Unbiased ratio estimators in stratified sampling. J. Am. Stat. Assoc. 56:70–87.

Peckham, C.J. 1989. Recent Trends in the International Tuna Market. NOAA/NMFS Tuna Newsletter, Issue 92.

Pereira, J. 1985. Composition Spécifique des bancs de Thons Pêchés à la Senne aux Acores. Pp. 395–400 in ICCAT, Collective Volume of Scientific Papers, Vol. 25, SCRS/85/82.

Perrin, W.F. 1968. The porpoise and the tuna. Sea Frontiers 14(3):166–174.

Perrin, W.F. 1969. Using porpoise to catch tuna. World Fishing 18:42–45.

Perrin, W.F. 1972. Variation and Taxonomy of Spotted and Spinner Porpoises (genus *Stenella*) of the Eastern Tropical Pacific and Hawaii. Ph.D. dissertation. University of California, Los Angeles. 490 pp.

Perrin, W.F. 1990. Subspecies of *Spenella longirostris* (Mammalia, Cetacea, Delphinidae). Biol. Soc. Wash. Proc. 102:453–463.

Perrin, W.F., and J.R. Hunter. 1972. Escape behavior of the Hawaiian spinner porpoise (*Stenella* of *S. longirostris*). Fish. Bull. 70:49–60.

Perrin, W.F., R.W. Warner, C.L. Fiscus, and D.B. Holts. 1973. Stomach contents of porpoise, *Stenella* spp., and yellowfin tuna, *Thunnus albacares*, in mixed species aggregation. Fish. Bull. 71:1077–1092.

Perrin, W.F., M.D. Scott, G.J. Walker, F.M. Ralston, and D.W.K. Au. 1983. Distribution of Four Dolphins (*Stenella* spp. and *Delphinus delphis*) in the Eastern Tropical Pacific, with an Annotated Catalog of Data Sources. NOAA-TM-NMFS-SWFC-38. National Marine Fisheries Service, Southwest Fisheries Center, La Jolla, Calif. 65 pp.

Perrin, W.F., R.L. Brownell, Jr., and D.P. DeMaster, eds. 1984. Reproduction of Whales, Dolphins and Porpoises. Cambridge, England: International Whaling Commission. 495 pp.

Perrin, W.F., M.D. Scott, G.J. Walker, and V.L. Cass. 1985. Review of Geographical Stocks of Tropical Dolphins (*Stenella* spp. and *Delphinus delphis*) in the Eastern Pacific. NOAA Technical Report NMFS 28. National Marine Fisheries Service, National Oceanic and Atmospheric Administration, U.S. Department of Commerce, Washington, D.C.

Polacheck, T. 1987. Relative abundance, distribution and inter-specific relationship of cetacean schools in the eastern tropical Pacific. Mar. Mammal Sci. 3:54–77.

Potier, M., and F. Marsac. 1984. La Pêche Thonèire dans l'Océan Indien; Campagne Exploratoire d'une Flottible de Senneurs (1982–1983). Rapp. Sci. Mission Orstom Seychelles 4. 73 pp.

Powers, J.E., R.W. Butler, J.G. Jennings, R. McLain, C.B. Peters, and J. DeBeer. 1979. Summary of Research Results from the Fourth Cruise of the Dedicated Vessel, 12 Sept.–31 Oct. 1978. Southwest Fisheries Center Administrative Report No. LJ-79–14. National Marine Fisheries Service, Southwest Fisheries Center, La Jolla, Calif. 42 pp.

Prado, J. 1988. Trawling for albacore. INFOFISH Int. (Kuala Lumpur). April:50.

Pryor, K.W., and I. Kang. 1980. Social Behavior and School Structure in Pelagic Porpoises, *Stenella attenuata* and *S. longirostris* during Purse Seining for Tuna. Southwest Fisheries Center Administrative Report No. LJ-80–11C. National Marine Fisheries Service, Southwest Fisheries Center, La Jolla, Calif.

Ralston, F., ed. 1977. A Workshop to Assess Research Related to the Tuna/porpoise Problem, February 28 and March 1–2. Southwest Fisheries Center Administrative Report No. LJ-77–15. National Marine Fisheries Service, Southwest Fisheries Center, La Jolla, Calif. 119 pp., 6 appendices.

Rao, J.N.K. 1969. Ratio and regression estimators. In New Developments in Survey Sampling, N.L. Johnson and H. Smith, Jr., eds. New York: Wiley Interscience. 732 pp.

Reilly, S.B., and J. Barlow. 1986. Rates of increase in dolphin population size. Fish. Bull. 84:527–533.

Robins, C.R., R.M. Bailey, C.E. Bond, J.R. Brooker, E.A. Lachner, R.N. Lea, and W.B. Scott. 1991. Common and Scientific Names of Fishes from the United States and Canada, 5th edition. Special Publ. No. 20. Bethesda, Md.: American Fisheries Society.

Sainsbury, J.C. 1971. Commercial Fishing Methods. Surrey, England: Fishing News (Books) Ltd.

Sakagawa, G.T. 1991. Are U.S. regulations on tuna-dolphin fishing driving U.S. seiners to foreign-flag registry? North Am. J. Fish. Manage. 11:241-252.

Schaefer, K.M. 1989. Morphometric analysis of yellowfin tuna, *Thunnus albacares*, from the eastern Pacific Ocean. Inter-Am. Trop. Tuna Comm., Bull. 19:389–427.

Scott, M.D. 1991. The Size and Structure of Pelagic Dolphin Herds. Ph.D. dissertation. University of California, Los Angeles. 185 pp.

Scott, M.D., and P.C. Wussow. 1983. Movements of a Hawaiian spotted dolphin. Pp. 353–364 in Proceedings of Fourth International Wildlife Biotelemetry Conference, D. Pincock, ed. Halifax, Nova Scotia.

Sexton, S.N., R.S. Holt, and D.P. DeMaster. In press. Investigating parameters affecting relative estimates in dolphin abundance in the eastern tropical Pacific from research vessel surveys in 1986, 1987, and 1988. IWC Doc. SC/42/SM43. Rep. Int. Whal. Comm.

Sharp, G.D., and A.E. Dizon, eds. 1978. The Physiological Ecology of Tunas. New York: Academic Press.

Simmons, D.C. 1968. Purse seining off Africa's West Coast. Comm. Fish. Rev. March:21–22.

Smith, T.D., ed. 1979. Report of the Status of Porpoise Stocks Workshop, August 27–31, 1979. Southwest Fisheries Center Administrative Report No. LJ-79–41. National Marine Fisheries Service, Southwest Fisheries Center, La Jolla, Calif.

Smith, T.D. 1983. Changes in the size of three dolphin (*Stenella* spp.) populations in the eastern tropical Pacific. Fish. Bull. 81:1–13.

Smith, T.D., and N.C.H. Lo. 1983. Some Data on Dolphin Mortality in the Eastern Tropical Pacific Tuna Purse Seine Fishery Prior to 1970. NOAA Tech. Memo. NOAA-TM-NMFS-SWFC-34. National Marine Fisheries Service, Southwest Fisheries Center, La Jolla, Calif. 26 pp.

Stretta, J-M., and M. Slepoukha. 1986. Analyse des facteurs biotiques et abiotiques associs aux Bancs de Thons. Pp. 161–169 in Proceedings of the ICCAT Conference on the International Skipjack Year Program, P.E.K.

Symons, P., M. Miyake, and G.T. Sakagawa, eds. International Commission for the Conservation of Atlantic Tunas, Madrid.

Stuntz, W.E. 1977. Report on the tuna-porpoise cruise to the Marine Mammal Commission. Pp. 33–37 in hearings before the Subcommittee on Fisheries and Wild Life Conservation and the Environment of the Committee Merchant Marine and Fisheries. House of Representatives, 95th Congress, First Session on Marine Mammal Oversight, February 17, Serial No. 95–8. Washington, D.C.: U.S. Government Printing Office.

Stuntz, W.E. 1981. The Tuna-Dolphin Bond: A Discussion of Current Hypotheses. Southwest Fisheries Center Administrative Report No. LJ-81–19. National Marine Fisheries Service, Southwest Fisheries Center, La Jolla, Calif.

Summers, C.B. 1990. Understanding current flow may help a trawl fish better. National Fisherman, June.

Sund, P.N., M. Blackburn, and F. Williams. 1981. Tunas and their environment in the Pacific Ocean: A review. Oceanog. Mar. Biol. Ann. Rev. 19:443-512.

Suzuki, Z., P.K. Tomlinson, and M. Honma. 1978. Population structure of Pacific yellowfin tuna, *Thunnus albacares*, from the eastern Pacific Ocean. Inter-Am. Trop. Tuna Comm., Bull. 17:273–441.

Thomas, J.A., and W.E. Evans. 1982. Final Report on Acoustic Detection of Tuna/porpoise Using a Towed Array. Tech. Rep. 82–138. Hubbs Sea World Research Institute, San Diego, Calif. 48 pp.

Tin, M. 1965. Comparison of some ratio estimators. J. Am. Stat. Assoc. 60:294–307.

U.S. International Trade Commission. 1990. Executive Summary. P. xiv in Tuna: Competitive Conditions Affecting the U.S. and European Tuna Industries in Domestic and Foreign Markets. ITC Publ. No. 2339. Washington, D.C.: U.S. International Trade Commission.

Van Dyke, J., and S. Heftel. 1981. The Management of Tuna in the Pacific: An Analysis of the South Pacific Forum Fisheries Agency. University of Hawaii Law Review 3(1).

Wahlen, B.E. 1986. Incidental dolphin mortality in the eastern tropical Pacific tuna fishery, 1973 through 1978. Fish. Bull. 84:559–569.

Waring, G.T., P. Gerrior, P.M. Payne, B.L. Parry, and J.R. Nicolas. 1990. Incidental take of marine mammals in foreign fishery activities off the northeast United States, 1977–88. Fish. Bull. 88:347–360.

Wild, A. 1986. Growth of yellowfin tuna, *Thunnus albacares*, in the eastern Pacific Ocean based on otolith increments. Inter-Am. Trop. Tuna Comm., Bull. 18:423–482.

Wild, A., and T.J. Foreman. 1980. The relationship between otolith increments and time for yellowfin and skipjack tunas marked with tetracycline. Inter-Am. Trop. Tuna Comm., Bull. 17:507–560.

APPENDIX

1

Operational regulations for tuna fishermen from the Code of Federal Regulations. *October 1990.*

Subpart C—General Exceptions

§ 216.21 Actions permitted by international treaty, convention, or agreement.

The Act and these regulations shall not apply to the extent that they are inconsistent with the provisions of any international treaty, convention or agreement, or any statute implementing the same relating to the taking or importation of marine mammals or marine mammal products, which was existing and in force prior to December 21, 1972, and to which the United States was a party. Specifically, the regulations in subpart B of this part and the provisions of the Act shall not apply to activities carried out pursuant to the Interim Convention on the Conservation of North Pacific Fur Seals signed at Washington on February 9, 1957, and the Fur Seal Act of 1966, 16 U.S.C. 1151 through 1187, as in each case, from time to time amended.

§ 216.22 Taking by State or local government officials.

(a) A State or local government official or employee may take a marine mammal in the normal course of his duties as an official or employee, and no permit shall be required, if such taking:

(1) Is accomplished in a humane manner;

(2) Is for the protection or welfare of such mammal or for the protection of the public health or welfare; and

(3) Includes steps designed to insure return of such mammal, if not killed in the course of such taking, to its natural habitat. In addition, any such official or employee may, incidental to such taking, possess and transport, but not sell or offer for sale, such mammal and use any port, harbor, or other place under the jurisdiction of the United States. All steps reasonably practicable under the circumstances shall be taken by any such employee or official to prevent injury or death to the marine mammal as the result of such taking. Where the marine mammal in question is injured or sick, it shall be permissible to place it in temporary captivity until such time as it is able to be returned to its natural habitat. It shall be permissible to dispose of a carcass of a marine mammal taken in accordance with this subsection whether the animal is dead at the time of taking or dies subsequent thereto.

(b) Each taking permitted under this section shall be included in a written report to be submitted to the Secretary every six months beginning December 31, 1973. Unless otherwise permitted by the Secretary, the report shall contain a description of:

(1) The animal involved;

(2) The circumstances requiring the taking;

(3) The method of taking;

(4) The name and official position of the State official or employee involved;

(5) The disposition of the animal, including in cases where the animal has

been retained in captivity, a description of the place and means of confinement and the measures taken for its maintenance and care; and

(6) Such other information as the Secretary may require.

§ 216.23 Native exceptions.

(a) *Taking.* Notwithstanding the prohibitions of subpart B of this part 216, but subject to the restrictions contained in this section, any Indian, Aleut, or Eskimo who resides on the coast of the North Pacific Ocean or the Arctic Ocean may take any marine mammal without a permit, if such taking is:

(1) By Alaskan Natives who reside in Alaska for subsistence, or

(2) For purposes of creating and selling authentic native articles of handicraft and clothing, and

(3) In each case, not accomplished in a wasteful manner.

(b) *Restrictions.* (1) No marine mammal taken for subsistence may be sold or otherwise transferred to any person other than an Alaskan Native or delivered, carried, transported, or shipped in interstate or foreign commerce, unless:

(i) It is being sent by an Alaskan Native directly or through a registered agent to a tannery registered under paragraph (c) of this section for the purpose of processing, and will be returned directly or through a registered agent to the Alaskan Native; or

(ii) It is sold or transferred to a registered agent in Alaska for resale or transfer to an Alaskan Native; or

(iii) It is an edible portion and it is sold in an Alaskan Native village or town.

(2) No marine mammal taken for purposes of creating and selling authentic native articles of handicraft and clothing may be sold or otherwise transferred to any person other than an Indian, Aleut or Eskimo, or delivered, carried, transported or shipped in interstate or foreign commerce, unless:

(i) It is being sent by an Indian, Aleut or Eskimo directly or through a registered agent to a tannery registered under paragraph (c) of this section for the purpose of processing, and will be returned directly or through a registered agent to the Indian, Aleut or Eskimo; or

(ii) It is sold or transferred to a registered agent for resale or transfer to an Indian, Aleut, or Eskimo; or

(iii) It has first been transformed into an authentic native article of handicraft or clothing; or

(iv) It is an edible portion and sold (A) in an Alaskan Native village or town, or (B) to an Alaskan Native for his consumption.

(c) Any tannery, or person who wishes to act as an agent, within the jurisdiction of the United States may apply to the Director, National Marine Fisheries Service, U.S. Department of Commerce, Washington, DC 20235, for registration as a tannery or an agent which may possess and process marine mammal products for Indians, Aleuts, or Eskimos. The application shall include the following information:

(i) The name and address of the applicant;

(ii) A description of the applicant's procedures for receiving, storing, processing, and shipping materials;

(iii) A proposal for a system of book-keeping and/or inventory segregation by which the applicant could maintain accurate records of marine mammals received from Indians, Aleuts, or Eskimos pursuant to this section;

(iv) Such other information as the Secretary may request;

(v) A certification in the following language:

I hereby certify that the foregoing information is complete, true and correct to the best of my knowledge and belief. I understand that this information is submitted for the purpose of obtaining the benefit of an exception under the Marine Mammal Protection Act of 1972 (16 U.S.C. 1361 through 1407) and regulations promulgated thereunder, and that any false statement may subject me to the criminal penalties of 18 U.S.C. 1001, or to penalties under the Marine Mammal Protection Act of 1972.

(vi) The signature of the applicant.

The sufficiency of the application shall be determined by the Secretary, and in that connection, he may waive any requirement for information, or require any elaboration or further information deemed necessary. The registration of a tannery or other agent shall be subject to such conditions as the Secretary prescribes, which may include, but are not limited to, provisions regarding records, inventory segregation, reports, and inspection. The Secretary may charge a reasonable fee for processing such applications, including an appropriate apportionment of overhead and administrative expenses of the Department of Commerce.

(d) Notwithstanding the preceding provisions of this section, whenever, under the Act, the Secretary determines any species of stock of marine mammals to be depleted, he may prescribe regulations pursuant to section 103 of the Act upon the taking of such marine animals by any Indian, Aleut, or Eskimo and, after promulgation of such regulations, all takings of such marine mammals shall conform to such regulations.

§ 216.24 **Taking and related acts incidental to commercial fishing operations.**

NOTE TO § 216.24: The provisions of 50 CFR part 229, rather than § 216.24, will govern the incidental taking of marine mammals in the course of commercial fishing operations by persons using vessels of the United States, other than vessels used in the eastern tropical Pacific yellowfin tuna purse seine fishery, and vessels which have valid fishing permits issued in accordance with section 204(b) of the Magnuson Fishery Conservation and Management Act (16 U.S.C. 1824(b)) for the period from November 23, 1988, through October 1, 1993. Other commercial fisheries remain subject to regulations under § 216.24.

(a)(1) No marine mammals may be taken in the course of a commercial fishing operation unless: The taking constitutes an incidental catch as defined in § 216.3, a general permit and certificate(s) of inclusion have been obtained in accordance with these regulations and such taking is not in violation of such permit, certificate(s), and regulations.

(2) A vessel on a commercial fishing trip involving the utilization of purse seines to capture yellowfin tuna which is not operating under a category two general permit and certificates of inclusion, and which during any part of its fishing trip is in the Pacific Ocean area described in the General Permit

for gear Category 2 operations, must not carry more than two speedboats.

(3) Upon written request in advance of entering the General Permit area, the limitation in (a)(2) may be waived by the Regional Director of the Southwest Region for the purpose of allowing transit through the General Permit area. The waiver will provide in writing the terms and conditions under which the vessel must operate, including a requirement to report by radio to the Regional Director the vessel's date of exit from or subsequent entry to the permit area, in order to transit the area with more than two speedboats.

(b) *General permits.* (1) General permits to allow the taking of marine mammals, except those for which taking is prohibited under the Endangered Species Act of 1973, in connection with commercial fishing operations will be issued to persons using fishing gear in any one of the following categories:

(i) *Category 1: Towed or dragged gear.* Commercial fishing operations utilizing towed or dragged gear such as bottom otter trawls, bottom pair trawls, multi-rig trawls, and dredging gear;

(ii) *Category 2: Encircling gear, pursue seining involving the intentional taking of marine mammals.* Commercial fishing operations utilizing purse seines to capture tuna by international encircling marine mammals. Only vessels that meet the fishing gear and equipment requirements contained in § 216.24(d)(2)(iv) of these regulations may be included in this category;

(iii) *Category 3: Encircling gear, pursue seining not involving the international taking of marine mammals.* Commercial fishing operations utilizing pursue seining, which do not intentionally encircle marine mammals;

(iv) *Category 4: Stationary gear.* Commercial fishing operations utilizing stationary gear such as traps, pots, weirs, and pound nets;

(v) *Category 5: Other gear.* Commerical fishing operations utilizing trolling, gill nets, hook and line gear, and any gear not classified under paragraph (b)(1)(i), (b)(1)(ii), (b)(1)(iii), (b)(1)(iv), or (b)(1)(vi) of this section; and

(vi) *Category 6: Commercial passenger fishing vessel operation.* Commercial fishing operations from a commercial passenger fishing vessel for the purpose of active sportfishing as defined in § 216.3.

(2) Permits shall be issued as general permits to a class of fishermen using one of the general categories of gear set out above. Any member of such class may apply for a general permit on behalf of any members of the class. Subsequent to the granting of general permit, vessel owners, managing owners, or operators (as required) may make application to be included under the terms of a general permit by obtaining a certificate of inclusion. Applications for a general, permit shall contain:

(i) Name, address, and telephone number of the applicant. If the applicant is an organization or corporate entity, a copy of the corporate or organizational charter which sets forth the basis for application on behalf of a group of class of commercial fishermen must be included;

(ii) A description of permit for which application is being made;

(iii) A description of the fishing operations by which marine mammals are taken; and a statement explaining why the applicant cannot avoid taking marine mammals incidentally to commercial fishing operations;

(iv) The date when the general permit is requested to become effective;

(v) A list of the fish sought by persons requesting certificates under the general permit and the general areas of operations of their vessels.

(vi) A statement identifying the marine mammals and numbers of marine mammals which are expected to be taken under the general permit;

(vii) A statement by the applicant demonstrating that the requested taking of marine mammal species or stocks during commercial fishing operations is consistent with the purposes of the Act, and the applicable regulations established under section 103 of the Act.

(viii) A description of the procedures and techniques that will be utilized in order that takings under the permit will be consistent with the purposes and policies of the Act and these regulations; and

(ix) A certification, signed by the applicant, in the following language: I certify that the foregoing information is complete, true, and correct to the best of my knowledge and belief. I understand that this information is submitted for the purpose of obtaining a general permit under the Marine Mammal Protection Act of 1972 and regulations promulgated thereunder, and that any false statement may subject me to the criminal penalties of 18 U.S.C. 1001, or the penalties provided under the Marine Mammal Protection Act of 1972.

(3) The original and two copies of the application for general permit must be submitted to the Assistant Administrator. Applications should be received not less than 180 days prior to the date upon which the permit is to become effective. Assistance may be obtained by writing the Assistant Administrator or by calling the Office of Marine Mammals and Endangered Species, telephone number 202-634-7461.

(4) A general permit shall be valid for the time period indicated on the face of the permit. General permits may contain terms and conditions prescribed in accordance with section 104(b)(2) of the Act, 16 U.S.C. 1374(b)(2). General permits may be suspended, revoked, modified, or denied. Procedures governing permit sanctions or denials for reasons relating to enforcement are found at subpart D of 15 CFR part 904.

(5) The Assistant Administrator shall determine the adequacy and completeness of an application, and if found to be adequate and complete will promptly publish a notice of receipt of such application in the FEDERAL REGISTER. Interested parties will have thirty days from the date of publication in which to submit written comments with respect to the granting of such permit.

(6) If within thirty days after the date of publication of the FEDERAL REGISTER notice concerning receipt of an application for a general permit, any interested party or parties request a hearing on the application, the Assistant Administrator may within sixty days following the date of publication of the FEDERAL REGISTER notice afford such party or parties an opportunity for such a hearing. Any hearing held in connection with an application for a general permit shall be conducted in the same manner as hearings convened in connection with a scientific research or a public display permit application under § 216.33.

(7) There is no fee for filing an application for a general permit.

(c) *Certificates of inclusion—*(1) *Vessel certificates of inclusion.* The owner or managing owner of a vessel that participates in commercial fishing operations for which a general permit is required under this subpart shall be the holder of a valid vessel certificate of inclusion under that general permit. Such certificates shall not be transferable and shall be renewed annually. Provided advance written notice is given, a vessel certificate holder may surrender his certificate to the Regional Office from which the certificate was issued. However, once surrendered the certificate shall not be returned nor shall a new certificate be issued before the end of the calendar year. This provision shall not apply when a change of vessel ownership occurs.

(2) *Operator's certificate of inclusion.* The person in charge of and actually controlling fishing operations (after this referred to as the operator) on a vessel engaged in commercial fishing operations for which a Category 2 or Category 6 general permit is required under this subpart, must be the holder of a valid operator's certificate of inclusion. These certificates are not transferable and will be valid only on a vessel having a valid vessel certificate of inclusion for the same category. In order to receive a certification of inclusion, the operator must have satisfactorily completed required training. An operator's certificate of inclusion must be renewed annually.

(3) A vessel certificate issued pursuant to paragraph (c)(1) of this section shall be aboard the vessel while it is engaged in fishing operations and the operator's certificate issued pursuant to paragraph (c)(2) of this section

shall be in the possession of the operator to whom it was issued. Certificates shall be shown upon request to an enforcement agent or other designated agent of the National Marine Fisheries Service. However, vessels and operators at sea on a fishing trip on the expiration date of their certificate of inclusion, to whom or to which a certificate of inclusion for the next year has been issued, may take marine mammals under the terms of the new certificate.

The vessel owners or operators are obligated to obtain physically or to place the new certificate aboard, as appropriate, when the vessel next returns to port.

(4) Application(s) for certificates of inclusion under paragraph (c)(1) of this section should be addressed as follows:

(i) Category 1, 3, 4, 5, and 6 applications:

(A) Owners or managing owners of vessels registered in Colorado, Idaho, Montana, North Dakota, Oregon, South Dakota, Utah, Washington, and Wyoming, should make application to the Regional Director, Northwest Region, National Marine Fisheries Service, 1700 Westlake Avenue, Seattle, Washington 98102.

(B) Owners or managing owners of vessels registered in Arizona, California, Hawaii, Nevada, and the territories of American Samoa, Guam, and the Trust Territory of the Pacific Islands should make application to the Regional Director, Southwest Region, National Marine Fisheries Service, 300 South Ferry Street, Terminal Island, California 90731.

(C) Owners or managing owners of vessels registered in Alaska should make application to the Regional Director, Alaska Region, National Marine Fisheries Service, P.O. Box 1668, Juneau, Alaska 99802.

(D) Owners or managing owners of vessels registered in Connecticut, Delaware, District of Columbia, Illinois, Indiana, Maine, Maryland, Massachusetts, Michigan, Minnesota, New Hampshire, New Jersey, New York, Ohio, Pennsylvania, Rhode Island, Vermont, Virginia, West Virginia, and Wisconsin should make application to the Regional Director, Northeast Region, National Marine Fisheries Service, 14 Elm Street, Federal Building, Gloucester, Massachusetts 01930.

(E) Owners or managing owners of vessels registered in Alabama, Arkansas, Florida, Georgia, Iowa, Kansas, Kentucky, Louisiana, Mississippi, Missouri, Nebraska, New Mexico, North Carolina, Oklahoma, Puerto Rico, South Carolina, Tennessee, Texas, and Virgin Islands, should make application to the Regional Director, National Marine Fisheries Service, Southeast Region, 9450 Gandy Boulevard North, Duval Building, St. Petersburg, Florida 33702.

(ii) Category 2 applications: Owners or managing owners of purse seine vessels in this category shall make application to the field office, National Marine Fisheries Service, Southwest Region, 1520 State Street, suite 200, San Diego, California 92101.

(5) Applications for vessel certificates of inclusion under paragraph (c)(1) of this section shall contain:

(i) The name of the vessel which is to appear on the certificate(s) of inclusion;

(ii) The category of the general permit under which the applicant wishes to be included;

(iii) The species of fish sought and general area of operations;

(iv) The identity of State and local commercial fishing licenses, if applicable, under which vessel operations are conducted, and dates of expiration;

(v) The name of the operator and date of training, if applicable; and

(vi) The name and signature of the applicant, whether owner or managing owner, address, and if applicable, the organization acting on behalf of the vessel.

(6) *Fees.* (i) Applications for certificates of inclusion under paragraph (c)(1) of this section shall contain a payment for each vessel named in the application in accordance with the following schedule:

(A) Categories 1: Towed or dragged gear; 3: Encircling gear, purse seining not involving the intentional taking of marine mammals; 4: Stationary gear; 5: Other gear; and 6: Commercial passenger fishing vessel—$10.00.

(B) Category 2: Encircling gear, purse seining involving the intentional taking of marine mammals—$200.00.

(ii) Except as provided herein, vessel owners or managing owners desiring a vessel certificate of inclusion under more than one category of the general permit will not be required to pay a full fee for each certificate. After the initial fee for a certificate is paid for each vessel, additional certificates will be issued for a fee of $.50 (fifty cents) each. However, every application for a vessel certificate under Category 2 shall contain the full fee.

(iii) Notwithstanding the provisions of paragraph (c)(6)(i) of this section, an applicant whose income is below Federal poverty guidelines may, upon showing in his application that his income is below such guidelines, be issued a certificate under the following schedule of fee payment:

(A) Categories, 1: Towed or dragged gear; 3: Encircling gear, purse seining not involving the intentional taking of marine mammals; 4: Stationary gear; 5: Other gear; and 6: Commercial passenger fishing vessels—$1.00.

(B) Category 2: Encircling gear, purse seining involving the intentional taking of marine mammals—$20.00.

(iv) A fee is not required for an operator's certificate of inclusion.

(v) The Assistant Administrator may change the amount of these required fees at any time he determines a different payment to be reasonable, and said change shall be accomplished by publication in the FEDERAL REGISTER of the new fee schedule.

(7) The Regional Office receiving applications for certificates of inclusion from vessel owners, managing owners, or operators shall determine the adequacy and completeness of such applications, and upon its determination that such applications are adequate and complete, it shall approve such applications and issue the certificate(s).

(8) Failure to comply with provisions of the general permit, certificate, or these regulations may lead to suspension, revocation, modification, or denial of a certificate of inclusion. It may also subject the certificate holder, vessel, vessel owner, operator, or master to the penalties provided

under the Act. Procedures governing permit sanctions and denials are found at subpart D of 15 CFR part 904.

(9) By using an operator or vessel certificate of inclusion under the Category 2 general permit, the certificate holder authorizes the release to the National Marine Fisheries Service of all data collected by observers aboard purse seine vessels during fishing trips under the Inter-American Tropical Tuna Commission observer program or any other international observer program in which the United States may participate. The certificate holder must furnish the international observer program all release forms required to provide the observer data to the National Marine Fisheries Service. Data obtained under such releases will be used for the same purposes as data collected directly by observers placed by the National Marine Fisheries Service and will be subject to the same standards of confidentiality.

(d) Terms and conditions of certificates under general permits shall include, but are not limited to the following:

(1) *Towed or dragged gear.* (i) A certificate holder may take marine mammals so long as such taking is an incidental occurrence in the course of normal commercial fishing operations. Marine mammals taken incidental to commercial fishing operations shall be immediately returned to the environment where captured without further injury.

(ii) A certificate holder may take such steps as are necessary to protect his catch, gear, or person from depredation, damage, or personal injury without inflicting death or injury to any marine mammal.

(iii) Only after all means permitted by paragraph (d)(1)(ii) of this section have been taken to deter a marine mammal from depredating the catch, damaging the gear, or causing personal injury, may the certificate holder injure or kill the animal causing the depredation or immediate personal injury; however, in no event shall a certificate holder kill or injure an Atlantic bottlenosed dolphin, *Tursiops truncatus,* under the provisions of this paragraph. A certificate holder shall not injure or kill any animal permitted to be killed or injured under this paragraph unless the infliction of such damage is substantial and immediate and is actually being caused at the time such steps are taken. In all cases, the burden is on the certificate holder to fully report and demonstrate that the animal was causing substantial and immediate damage or about to cause personal injury and that all possible steps to protect against such damage or injury as permitted by paragraph (d)(1)(ii) of this section were taken and that such attempts failed.

(iv) Marine mammals taken in the course of commercial fishing operations shall be subject to the provisions of § 216.3 with respect to "Incidental catch," and may not be retained except where a specific permit has been obtained authorizing the retention.

(v) All certificate holders shall maintain logs of incidental take of marine mammals in such form as prescribed by the Assistant Administrator. All deaths or injuries to marine mammals occurring in the course of commercial fishing operations under the conditions of a general permit shall be immediately recorded in the log and reported in writing to the Regional Director to whom the certificate application was made, or to an enforcement agent or other designated agent of the National Marine Fisheries Service, at the earliest opportunity, but no later than five days after such occurrence, except that if a vessel at sea returns to port later than five days after such occurrence then it shall be reported within 48 hours after arrival in any port. Reports must include:

(A) The location, time, and date of the death or injury;

(B) The identity and number of marine mammals killed or injured; and

(C) A description of the circumstances which led up to and caused the death or injury.

(2) *Encircling gear, purse seining involving the intentional taking of marine mammals*—(i) *Quotas:*

(A) A certificated vessel may take marine mammals so long as the taking is an incidental occurrence in the course of normal commercial tuna purse seine fishing operations, and the

fishing operations are under the immediate direction of a person who is the holder of a valid operator's certificate of inclusion; except that a vessel shall not encircle either:

(1) Pure schools of any species of dolphin except the offshore spotted dolphin (*Stenella attenuata*) stock, the striped dolphin (*Stenella coeruleoalba*) species, and the common dolphin (*Dephinus delphis*) species; or

(2) Any other species or stock or marine mammals that does not have an allowable take as listed below or whose allowable take has been exceeded. The numbers of marine mammals that may be taken during each calendar year by U.S. vessels in the course of commercial fishing operations will be limited to:

Quotas for each calendar year		Take	Encircle-ment	Mortali-ty[1]
Species/ stock	Manage-ment unit			
Spotted dolphin.	(northern offshore)[2].	16,570,000	10,338,000	20,500
Do	(southern offshore).	4,605,000	2,873,000	5,697
Do	(coastal)	202,000	126,000	[3] 250
Spinner dolphin.	(eastern)	2,222,000	1,386,000	[3]2,750
Do	(northern whitebelly).	1,205,000	699,000	5,321
Do	(southern whitebel-ly)[4].	568,000	329,000	2,506
Common dolphin.	(northern tropical)[5].	723,000	450,000	1,890
Do	(central tropical).	2,619,000	845,000	8,112
Do	(southern tropical).	1,306,000	421,000	4,045
Striped dolphin.	(northern tropical).	28,000	21,000	429
Do	(central tropical).	118,000	89,000	1,822
Do	(southern tropical).	265,000	199,000	4,095

[1] The U.S. allowable mortality in any one year may not exceed 20,500.
[2] Fifty percent of replacement yield for the northern offshore spotted dolphin is 42,898; however, the maximum allowable mortality in any year is 20,500.
[3] Mortality level established by Pub. L. 98–364; not subject to flexible mortality schedule published in 46 FR 42068–42069 (August 19, 1981).
[4] Includes allowance for mixed species take.
[5] Includes Baja neritic dolphin stock.

(B) The incidental mortality of marine mammals permitted under the general permit for each category will be monitored according to the methodology published in the FEDERAL REGISTER. The Assistant Administrator shall determine on the basis of the evidence available to him the date upon which the allowable quotas will be reached or exceeded. Notice of the Assistant Administrator's determination shall be published in the FEDERAL REGISTER not less than seven days prior to the effective date.

(C) Except for the coastal spotted dolphin stock and the eastern spinner dolphin stock, if at the time the net skiff attached to the net is released from the vessel at the start of a set, and species or stocks that are prohibited from being taken are not reasonably observable, the fact that individuals of that species or stock are subsequently taken will not be cause for issuance of a notice of violation provided that all procedures required by the applicable regulations have been followed.

(D) The general permit is valid until surrendered by the permit holder or suspended or terminated by the Assistant Administrator provided the permittee and certificate holders under this part continue to use the best marine mammal safety techniques and equipment that are economically and technologically practicable. The Assistant Administrator may, upon receipt of new information which in his opinion is sufficient to require modification of the general permit or regulations, propose to modify such after consultation with the Marine Mammal Commission. These modifications must be consistent with and necessary to carry out the purposes of the Act. Any modifications proposed by the Assistant Administrator involving changes in the quotas will include the statements required by section 103(d) of the Act. Modifications will be proposed in the FEDERAL REGISTER and a public comment period will be allowed. At the request of any interested person within 15 days after publication of the proposed modification in the FEDERAL REGISTER, the Assistant Administrator may hold a public hearing to receive and evaluate evidence in those circumstances where he has determined it to be consistent with and necessary to carry out the purposes of the Act. Such request may be for a formal hearing on the record before an Administrative Law Judge. Within

10 days after receipt of the request for a public hearing, the Assistant Administrator will provide the requesting party or parties with his decision. If a request is denied, the Assistant Administrator will state the reasons for the denial. Within 10 days after receipt of a decision denying a request for a formal hearing, the requesting person may file a written notice of appeal with the Administrator. Based upon the evidence presented in the notice, the Administrator will render a decision within 20 days from receipt of the notice.

(ii) *General conditions:* (A) Marine mammals incidentally taken must be immediately returned to the environment where captured without further injury. The operators of purse seine vessels must take every precaution to refrain from causing or permitting incidental mortality or serious injury of marine mammals. Live marine mammals must not be brailed or hoisted onto the deck during ortza retrieval.

(B) Operators may take such steps as are necessary to protect their gear or person from damage or threat of personal injury. However, all marine mammals taken in the course of commercial fishing operations shall be subject to the definition of "incidental catch" in § 216.3 of this part and may not be retained except where a specific permit has been obtained authorizing the retention.

(C) The vessel certificate holder shall notify the field office, National Marine Fisheries Service, Southwest Region, 1520 State Street, suite 200, San Diego, California 92101, telephone 619-557-6540, of any change of vessel operator within at least 48 hours prior to departing on the next scheduled trip.

(iii) *Reporting requirements:* In accordance with § 216.24(f) of these regulations, the following specific reporting procedures shall be required:

(A) The vessel certificate holder of each certificated vessel, who has been notified via certified letter from the National Marine Fisheries Service that his vessel is required to carry an observer, shall notify the field office, Southwest Region, National Marine Fisheries Service, San Diego, California, telephone 714-293-6540 at least five (5) days in advance of the vessel's departure on a fishing voyage to allow for observer placement. After a fishing voyage is initiated, the vessel is obligated to carry an observer until the vessel returns to port and one of the following conditions is met:

(*1*) Unloads more than 400 tons of any species of tuna; or (*2*) unloads any amount of any species of tuna equivalent to one half of the vessel's carrying capacity; or (*3*) unloads its tuna catch after 40 days or more at sea from the date of departure. Further, the Regional Director, Southwest Region, may consider special circumstances for exemptions to this definition, provided written requests clearly describing the circumstances are received prior to the termination or the initiation of a fishing voyage. A response to the written request will be made by the Regional Director within five (5) days after receipt of the request. A vessel whose vessel certificate holder has failed to comply with the provisions of this section may not engage in fishing operations for which a general permit is required.

(B) Masters of all certificated vessels carrying National Marine Fisheries Service observers shall allow observers to report, in coded form, information by radio concerning the accumulated take of marine mammals and other observer collected data at such times as specified by the Regional Director, Southwest Region. Individual vessel names and coded information reported by radio by the National Marine Fisheries Service observers shall remain confidential unless their release is authorized in writing by the operator of the vessel.

(C) The Regional Director, Southwest Region, will provide to the public, periodic quota status reports summarizing the estimated incidental porpoise mortality by U.S. vessels of individual species and stock.

(iv) A vessel having a vessel certificate issued under paragraph (c)(1) may not engage in fishing operations for which a general permit is required unless it is equipped with a porpoise safety panel in its purse seine, and has and uses the other required gear, equipment, and procedures.

(A) *Class I and II Vessels:* For Class I purse seiners (400 short tons carrying capacity or less) and for Class II purse seiners (greater than 400 short tons carrying capacity, built before 1961), the porpoise safety panel must be a minimum of 100 fathoms in length (as measured before installation), except that the minimum length of the panel in nets deeper than 10 strips must be determined at a ratio of 10 fathoms in length for each strip that the net is deep. It must be installed so as to protect the perimeter of the backdown area. The perimeter of the backdown area is the length of the corkline which begins at the outboard end of the last bow bunch pulled and continues to at least two-thirds the distance from the backdown channel apex to the stern tiedown point. The porpoise safety panel must consist of small mesh webbing not to exceed 1¼" stretch mesh, extending from the corkline downward to a minimum depth equivalent to one strip of 100 meshes of 4¼" stretch mesh webbing. In addition, at least a 20-fathom length of corkline must be free from bunchlines at the apex of the backdown channel.

(B) *Class III Vessels:* For Class III purse seiners (greater than 400 short tons carrying capacity, built after 1960), the porpoise safety panel must be a minimum of 180 fathoms in length (as measured before installation), except that the minimum length of the panel in nets deeper than 18 strips must be determined in a ratio of 10 fathoms in length for each strip of net depth. It must be installed so as to protect the perimeter of the backdown area. The perimeter of the backdown area is the length of corkline which begins at the outboard end of the last bowbunch pulled and continues to at least two-thirds the distance from the backdown channel apex to the stern tiedown point. The porpoise safety panel must consist of small mesh webbing not to exceed 1¼" stretch mesh extending downward from the corkline and, if present, the base of the porpoise apron to a minimum depth equivalent to two strips of 100 meshes of 4¼" stretch mesh webbing. In addition, at least a 20-fathom length of corkline must be free from bunchlines at the apex of the backdown channel.

(C) *Porpoise safety panel markers:* Each end of the porpoise safety panel and porpoise apron shall be identified with an easily distinguishable marker.

(D) *Porpoise safety panel hand holds:* Throughout the length of the corkline under which the porpoise safety panel and porpoise apron are located, hand hold openings are to be secured so that the insertion of a 1⅜" diameter cylindrical-shaped object meets resistance.

(E) *Porpoise safety panel corkline hangings:* Throughout the length of the corkline under which the porpoise safety panel and porpoise apron are located, corkline hangings shall be inspected by the vessel operator following each trip. Hangings found to have loosened to the extent that a cylindrical object with a 1⅜" diameter will not meet resistance when inserted between the cork and corkline hangings, must be tightened so that a cylindrical object with a 1⅜" diameter cannot be inserted.

(F) *Speedboats:* Certificated vessels engaged in fishing operations involving setting on marine mammals shall carry a minimum of two speedboats in operating condition. All speedboats carried aboard purse seine vessels and in operating condition shall be rigged with towing bridles and towlines. Speedboat hoisting bridles shall not be substituted for towing bridles.

(G) *Raft:* A raft suitable to be used as a porpoise observation-and-rescue platform shall be carried on all certificated vessels.

(H) *Facemask and snorkel, or viewbox:* At least two facemasks and snorkels, or viewboxes, must be carried on all certificated vessels.

(I) *Lights:* All certificated vessels shall be equipped by July 1, 1986, with lights capable of producing a minimum of 140,000 lumens of output for use in darkness to ensure sufficient light to observe that procedures for porpoise release are carried out and to monitor incidental porpoise mortality.

(v) *Vessel inspection:* (A) *Annual:* At least once during each calendar year, purse seine nets and other gear and equipment required by these regulations shall be made available for in-spection by an authorized National Marine Fisheries Service Inspector as specified by the Regional Director, Southwest Region.

(B) *Reinspection:* Purse seine nets and other gear and equipment required by these regulations shall be made available for reinspection by an authorized National Marine Fisheries Service Inspector as specified by the Regional Director, Southwest Region. The vessel certificate holder shall notify the Fleet Assistance Section, , National Marine Fisheries Service, Southwest Region, 1520 State Street, suite 200, San Diego, California 92101, telephone 619–557–6540, of any net modification at least five (5) days prior to departure of the vessel on its next scheduled trip in order to determine whether a reinspection or trial set would be required.

(C) Upon failure to pass an inspection or reinspection, a vessel having a vessel certificate of inclusion issued under paragraph (c)(1) may not engage in fishing operations for which a general permit is required until the deficiencies in gear or equipment are corrected as required by an authorized National Marine Fisheries Service inspector.

(vi) *Operator training requirements.* All operators shall maintain proficiency sufficient to perform the procedures required herein, and must attend and satisfactorily complete a formal training session conducted under the auspices of the National Marine Fisheries Service in order to obtain their certificate of inclusion. At the training session an attendee shall be instructed concerning the provisions of the Marine Mammal Protection Act of 1972, the regulations promulgated pursuant to the Act, and the fishing gear and techniques which are required or will contribute to reducing serious injury and mortality of porpoise incidental to purse seining for tuna. Operators who have received a written certificate of satisfactory completion of training and who possess a current or previous calendar year certificate of inclusion will not be required to attend additional formal training sessions unless there are substantial changes in the Act, the regulations, or the required fishing gear and techniques. Additional training may be required for any operator who is found by the Regional Director, Southwest Region, to lack proficiency in the procedures required.

(vii) *Marine mammal release requirements:* All operators shall use the following procedures during all sets involving the incidental taking of marine mammals in association with the capture and landing of tuna.

(A) *Backdown procedure:* Backdown shall be performed following a purse seine set in which marine mammals are captured in the course of catching and landing tuna, and shall be continued until it is no longer possible to remove live marine mammals from the net by this procedure. Thereafter, other release procedures required shall be continued until all live animals have been released from the net.

(B) *Prohibited use of sharp or pointed instrument:* The use of a sharp or pointed instrument to remove any marine mammal from the net is prohibited.

(C) *Sundown sets prohibited.* On every set encircling porpoise, the backdown procedure must be completed and rolling of the net to sack-up must be begun before one-half hour after sundown, except as provided below. For the purpose of this section, "sundown" is defined as the time at which the upper edge of the sun disappears below the horizon or, if view of the sun is obscured, the local time of sunset calculated from tables developed by the U.S. Naval Observatory. A "sundown set" is a set in which the backdown procedure has not been completed and rolling the net to sack-up has not begun within one-half hour after sundown. Should a set extend beyond one-half hour after sundown, the operator must use the required marine mammal release procedures including the use of the high intensity lighting system.

(1) A certificated operator may obtain an initial waiver from this prohibition, for trips with an observer, by establishing to the satisfaction of the National Marine Fisheries Service (NMFS) Southwest Regional Director, based upon NMFS and Inter-American Tropical Tuna Commission (IATTC) observer records, that the operator's

average kill of marine mammals per ton of yellowfin tuna caught in sundown sets involving marine mammals was 0.154 mammals or fewer.

(i) The application must include the following:

(A) Name of the operator as it appears on the certificate of inclusion;

(B) The dates of all observed trips any part of which occurred since July 1, 1986 and observed trips before that date, if necessary to include a minimum of three observed sundown sets;

(C) Names of the vessels operated during those trips;

(D) The number of marine mammals killed and the number of tons of yellowfin tuna caught in sundown sets involving marine mammals;

(E) Detailed description of the circumstances that support any request that the mortality associated with a particular sundown set be excluded from consideration; and

(F) The operator's signature or the signature of an individual authorized by the operator to make the application in the operator's absence.

(ii) All sundown sets since July 1, 1986 will be considered for this determination, except that the Regional Director will exclude one sundown set from each twelve month period from the calculations of average kill if the operator establishes to the satisfaction of the Regional Director that the kill in that sundown set was due to an unforeseeable equipment malfunction that could not have been avoided by reasonable diligence in operating or maintaining the vessel.

(iii) An operator must have a minimum of five observed sundown sets for the Regional Director to consider in determining whether or not the operator qualifies for an exemption. If an operator does not have five observed sundown sets since July 1, 1986, the NMFS Southwest Regional Director will consider records from observed trips before that date, starting with the most recent observed trip during which a sundown set was made and reviewing as many trips as necessary to obtain at least five sundown sets for consideration.

(2) An operator fishing under an exemption from the sundown set prohibition must follow the marine

mammal release requirements, including the use of high intensity lights for sets that continue one-half hour past sundown.

(3) An operator exemption is valid for one calendar year only on trips carrying a NMFS or IATTC observer and expires on December 31, unless renewed by the Regional Director.

(4) An exemption will be reviewed annually between November 1 and December 15 and the exemption will not be renewed if the operator's average mortality in sundown sets during trips completed in the previous twelve month period ending November 1 exceeds the United States fleet's average mortality rate in daylight sets for all of the observed trips completed in the same period.

(5) An operator who is notified that his or her exemption will not be renewed, or who anticipates not getting renewed, may petition the Regional Director in writing to reinstate the exemption based on excluding from the calculations one set where an unforeseeable equipment malfunction caused mortality in a sundown set that could not have been avoided by reasonable diligence in operating or maintaining the vessel. The Regional Director will reinstate the exemption if the evidence supports excluding the set and if the resulting recalculation of the operator's performance meets the standard required by these regulations.

(D) *Porpoise Safety Panel:* During backdown, the porpoise safety panel must be positioned so that it protects the perimeter of the backdown area. The perimeter of the backdown area is the length of corkline which begins at the outboard end of the last bow bunch pulled and continues to at least two-thirds the distance from the backdown channel apex to the stern tiedown point. Any super apron must be positioned at the apex of the backdown channel.

(E) *Use of explosive devices:* The use of explosive devices is prohibited in all tuna purse seine operations that involve marine mammals.

(viii) *Experimental fishing operations:* The Regional Director, Southwest Region, may authorize experimental fishing operations and may waive, as appropriate, any require-

ments within § 216.24(d)(2), except quotas on the incidental kill of marine mammals and the prohibition on setting nets on pure schools of certain porpoise species.

(A) A vessel certificate holder may apply for an experimental fishing operation waiver by submitting the following information to the Southwest Regional Director no less than 90 days before the intended date the proposed operation is intended to begin·

(1) Name(s) of the vessel(s) and the vessel certificate holder(s) to participate;

(2) A statement of the specific vessel gear and equipment or procedural requirement to be exempted and why such an exemption is necessary to conduct the experiment;

(3) A description of how the proposed modification to the gear or procedures is expected to reduce incidental mortalities or serious injury of marine mammals;

(4) A description of the applicability of this modification to other purse seine vessels;

(5) Planned design, time, duration, and general area of the experimental operation;

(6) Name(s) of the certificated operator(s) of the vessel(s) during the experiment;

(7) A statement of the qualifications of the individual or company doing the analysis of the research.

(B) The Regional Director will acknowledge receipt of the application and, upon determining that it is complete, ·publish notice in the FEDERAL REGISTER summarizing the application, making the full application available for inspection and inviting comments for a minimum period of thirty days from the date of publication.

(C) The Regional Director, after considering the information identified in paragraph (d)(2)(viii)(A) of this section and the comments received, will deny the application giving the reasons for denial or issue a permit to conduct the experiment including restrictions and conditions as deemed appropriate.

(D) The permit for an experimental fishing operation will be valid only for the vessels and operators named in the permit, for the time period and areas specified, for trips carrying an observer assigned by the NMFS, and when all the terms and conditions of the permit are met.

(E) The Regional Director may suspend or revoke an experimental fishing permit by written notice to the permit holder if the terms and conditions of the permit or the provisions of the regulations are not followed, after providing an opportunity for the permit holder to discuss the proposed suspension or revocation.

(ix) *Operator Certificate of Inclusion Holder Performance Requirements.* (A) The certificate of inclusion of any operator who makes one or more purse seine sets on marine mammals resulting in an average kill-per-set for a fishing trip which exceeds 26.30 marine mammals is suspended. Such suspension shall be effective upon notification from the National Marine Fisheries Service Southwest Regional Director and shall be for a period of one year. If the operator exceeds the 26.30 marine mammals killed per set average for a subsequent trip within three years of reinstatement, the certificate is suspended. Such suspension shall be effective upon notification by the Regional Director and shall be for a period of one year. If the operator's average mortality rate exceeds 26.30 marine mammals kill-per-set on a subsequent trip within five years of the second reinstatement, the certificate is revoked. The revocation shall be effective upon notification by the Regional Director and shall be permanent. An operator who is subject to a suspension or revocation under this paragraph may petition the Regional Director to review the operator's marine mammal mortality history. The Regional Director may reinstate the operator's certificate if the operator demonstrates that the operator has not exceeded a kill-per-set of 3.89 marine mammals during any of the eight consecutive observed trips immediately preceding the trip which caused the suspension. However, that trip will be considered as a single trip exceeding a kill-per-set of 3.89 marine mammals and subject to the conditions described in paragraph (d)(2)(ix)(F) of this section. The Regional Director may exclude from the mortality calculation for a trip, those purse seine sets in which marine mammal mortality resulted from an unavoidable and unforeseeable equipment breakdown. The mortality rate calculated after exclusion of a set or sets under this paragraph will determine the action taken under this performance evaluation system.

(B) Fishing trips with five or fewer sets on marine mammals and an average kill-per-set less than or equal to 26.30 marine mammals are not subject to further action under the operator performance system. Such trips neither count as trips meeting the performance standard nor count as trips failing to meet the performance standard for the purpose of determining actions based on performance in consecutive fishing trips.

(C) Fishing trips with more than five sets on marine mammals resulting in an average kill-per-set of not greater than 26.30 marine mammals are subject to review under the operator performance system as follows:

(*1*) The operator's kill of marine mammals in purse seine sets on marine mammals will be determined from observer records.

(*2*) The kill-per-set will be determined by dividing the total kill of marine mammals by the number of sets involving marine mammals during the fishing trip.

(*3*) If the calculated kill-per-set for the trip is equal to or less than 3.89 marine mammals, the operator has met the performance standard and is not subject to further action under the performance system based on the current trip.

(*4*) If the calculated kill-per-set for the trip exceeds 3.89 marine mammals, the operator failed to meet the mortality performance standard and is subject to further action under the performance system.

(D) The Southwest Regional Director may exclude from the mortality calculation for a trip, those purse seine sets in which marine mammal mortality resulted from an unavoidable and unforeseeable equipment breakdown. Should exclusion of a set or sets cause the operator's performance to fall within the standard performance, that trip will not be counted as a trip for the purposes of the performance evaluation system.

(E) An operator shall not serve as a certificated operator until the Southwest Regional Director has determined under this subpart and notified the operator that the operator's marine mammal mortality rate performance met or failed to meet the applicable performance standard on the previous observed trip. The Southwest Regional Director will make the determination within five days (excluding Saturdays, Sundays and Federal holidays) after receiving the observer data from the trip.

(F) An operator whose average marine mammal mortality rate exceeds 3.89 kill-per-set for a trip must have observer data and other pertinent records reviewed by the Southwest Regional Director and the Porpoise Rescue Foundation for the purpose of determining the causes of higher than acceptable mortality, must participate in supplemental marine mammal safety training as ordered by the Southwest Regional Director and must comply with actions for reducing marine mammal mortality which may be ordered by the Southwest Regional Director. The operator must carry an observer on the next trip for which he serves as the certificated operator. If the Southwest Regional Director determines that the required training or other ordered action has not been completed satisfactorily or is refused, the Regional Director will suspend the operator's certificate of inclusion for one year.

(G) An operator whose average marine mammal mortality rate exceeds 3.89 marine mammals killed per set on two consecutive trips or on three trips ending within a period of twenty-four months or on three trips within eight consecutive trips must have observer data and other pertinent records reviewed by the Southwest Regional Director and the Porpoise Rescue Foundation for the purpose of determining the causes of higher than acceptable mortality, must participate in supplemental marine mammal safety training as ordered by the Southwest Regional Director and must comply with actions for reducing marine mammal mortali-

ty which may be ordered by the Southwest Regional Director. The operator must carry an expert fisherman (*i.e.*, an experienced vessel operator with a history of low dolphin mortality), if required to do so by the Southwest Regional Director, to assist in perfecting marine mammal safety techniques, and must also carry an observer on the next trip for which he serves as the certificated operator. The selection of the expert fisherman will be provided by the General Permit holder or the Porpoise Rescue Foundation and subject to the approval of the Southwest Regional Director. If the Southwest Regional Director determines that the required training or other ordered action has not been completed satisfactorily or is refused, the Regional Director will suspend the operator's certificate of inclusion for one year.

(H) The operator certificate of inclusion or an operator whose average marine mammal mortality rate exceeds 3.89 kill-per-set on three *consecutive* trips, or on any four trips (of which no more than two are consecutives) completed within a period of twenty-four months or on four trips (of which no more than two are consecutive) within eight consecutively observed trips, is suspended upon notification to the operator from the Regional Director.

(I) Following a suspension and a reinstatement of a certification of inclusion, the operator certificate of inclusion is suspended for any operator whose average marine mammal mortality rate exceeds 3.89 marine mammals killed per set on any subsequent trip as required under the criteria for a suspension established in paragraph (d)(2)(ix)(H) of this section. Under this paragraph, trips completed by the operator prior to suspension will be carried over and counted along with trips completed subsequent to the suspension. Such suspension shall be effective upon notification from the NMFS Southwest Regional Director and shall be for a period of one year. For purposes of this paragraph only, each suspension under paragraph (d)(2)(ix)(A) of this section will be considered equivalent to and counted as three consecutive trips exceeding the trip kill rate of 3.89 marine mammals killed per set.

(J) An operator may appeal suspension of revocation of a certificate of inclusion under paragraphs (d)(2)(ix)(A), (d)(2)(ix)(H), or (d)(2)(ix)(I) of this section to the Assistant Administrator. Appeals must be filed in writing within 30 days of suspension or revocation and must contain a statement setting forth the basis for the appeal. Appeals must be filed with the Regional Director, Southwest Region, NMFS. The appeal may be presented at the option of the operator at a hearing before a person appointed by the Assistant Administrator to hear the appeal. The Assistant Administrator will determine, based upon the record, including any record developed at a hearing, if the suspension or revocation is supported under the criteria set forth in these regulations. The decision of the Assistant Administrator will be the final decision of the Department of Commerce.

(K) An operator must carry an observer on the operator's first trip after a suspension under this performance system has expired. An operator must also participate in supplemental marine mammal safety training and comply with actions for reducing marine mammal mortality as ordered by the Southwest Regional Director before making another trip as a certified operator.

(L) A person obtaining an operator certificate of inclusion for the first time must carry an observer on the operator's first trip.

(3) *Encircling gear, purse seining not involving the intentional taking of marine mammals.* (i) A certificate holder may take marine mammals so long as such taking is an incidental occurrence in the course of normal commercial fishing operations. Marine mammals taken incidental to commercial fishing operations shall be immediately returned to the environment where captured without further injury.

(ii) A certificate holder may take such steps as are necessary to protect his catch, gear, or person from depredation, damage or personal injury without inflicting death or injury to any marine mammal.

(iii) Only after all means permitted by paragraph (d)(3)(ii) of this section have been taken to deter a marine mammal from depredating the catch, damaging the gear, or causing personal injury, may the certificate holder injure or kill the animal causing the depredation or immediate damage, or about to cause immediate personal injury; however, in no event shall a certificate holder kill or injure an Atlantic bottlenosed dolphin, *Tursiops truncatus,* under the provisions of this paragraph. A certificate holder shall not injure or kill any animal permitted to be killed or injured under this paragraph unless the infliction of such damage is substantial and immediate and is actually being caused at the time such steps are taken. In all cases, the burden is on the certificate holder to report fully and demonstrate that the animal was causing substantial and immediate damage or about to cause personal injury and that all possible steps to protect against such damage or injury as permitted by paragraph (d)(3)(ii) of this section were taken and that such attempts failed.

(iv) Marine mammals taken in the course of commercial fishing operations shall be subject to the provisions of § 216.3 with respect to "Incidental catch," and may be retained except where a specific permit has been obtained authorizing the retention.

(v) All certificate holders shall maintain logs of incidental take of marine mammals in such form as prescribed by the Assistant Administrator. All deaths or injuries to marine mammals occurring in the course of commercial fishing operations under the conditions of a general permit shall be immediately recorded in the log and reported in writing to the Regional Director, National Marine Fisheries Service, where a certificate application was made, or to an enforcement agent or other designated agent of the National Marine Fisheries Service, at the earliest opportunity but no later than five days after such occurrence, except that if a vessel at sea returns to port later than five days after such occurrence, then it shall be reported within forty-eight hours after arrival in port. Reports must include:

(A) The location, time, and date of the death or injury;

(B) The identity and number of marine mammals killed or injured; and

(C) A description of the circumstances which led up to and caused the death or injury.

(4) *Stationary gear.* (i) A certificate holder may take marine mammals so long as such taking is an incidental occurrence in the course of normal commercial fishing operations. Marine mammals taken incidental to commercial fishing operations shall be immediately returned to the environment where captured without further injury.

(ii) A certificate holder may take such steps as are necessary to protect his catch, gear, or person from depredation, damage or personal injury without inflicting death or injury to any marine mammal.

(iii) Only after all means permitted by paragraph (d)(4)(ii) of this section have been taken to deter a marine mammal from depredating the catch, damaging the gear, or causing personal injury, may the certificate holder injure or kill the animal causing the depredation or immediate damage, or about to cause immediate personal injury; however, in no event shall a certificate holder kill or injure an Atlantic bottlenosed dolphin, *Tursiops truncatus,* under the provisions of this paragraph. A certificate holder shall not injure or kill any animal permitted to be killed or injured under this paragraph unless the infliction of such damage is substantial and immediate and is actually being caused at the time such steps are taken. In all cases, the burden is on the certificate holder to report fully and demonstrate that the animal was causing substantial and immediate damage or about to cause personal injury and that all possible steps to protect against such damage or injury as permitted by paragraph (ii) were taken and that such attempts failed.

(iv) Marine mammals taken in the course of commercial fishing operations shall be subject to the provisions of § 216.3 with respect to "Inci-

dental catch," and may not be retained except where a specific permit has been obtained authorizing the retention.

(v) All certificate holders shall maintain logs of incidental take of marine mammals in such form as prescribed by the Assistant Administrator. All deaths or injuries to marine mammals occurring in the course of commercial fishing operations under the conditions of a general permit shall be immediately recorded in the log and reported in writing to the Regional Director, National Marine Fisheries Service, where a certificate application was made, or to an enforcement agent or other designated agent of the National Marine Fisheries Service, at the earliest opportunity but no later than five days after such occurrence, except that if a vessel at sea returns to port later than five days after such occurrence, then it shall be reported within forty-eight hours after arrival in port. Reports must include:

(A) The location time, and date of the death or injury;

(B) The identity and number of marine mammals killed or injured; and

(C) A description of the circumstances which led up to and caused the death or injury.

(5) *Other gear.* (i) A certificate holder may take marine mammals so long as such taking is an incidental occurrence in the course of normal commercial fishing operations. Marine mammals taken incidental to commercial fishing operations shall be immediately returned to the environment where captured without further injury.

(ii) A certificate holder may take such steps as are necessary to protect his catch, gear, or person from depredation, damage or personal injury without inflicting death or injury to any marine mammal.

(iii) Only after all means permitted by paragraph (d)(5)(ii) of this section have been taken to deter a marine mammal from depredating the catch, damaging the gear, or causing personal injury, may the certificate holder injure or kill the animal causing the depredation or immediate damage, or about to cause immediate personal injury; however, in no event shall a certificate holder kill or injure an Atlantic bottlenosed dolphin, *Tursiops truncatus,* under the provisions of this paragraph. A certificate holder shall not injure or kill any animal permitted to be killed or injured under this paragraph unless the infliction of such damage is substantial and immediate and is actually being caused at the time such steps are taken. In all cases, the burden is on the certificate holder to report fully and demonstrate that the animal was causing substantial and immediate damage or about to cause personal injury and that all possible steps to protect against such damage or injury as permitted by paragraph (d)(5)(ii) of this section were taken and that such attempts failed.

(iv) Marine mammals taken in the course of commercial fishing operations shall be subject to the provisions of § 216.3 with respect to "Incidental catch," and may not be retained except where a specific permit has been obtained authorizing the retention.

(v) All certificate holders shall maintain logs of incidental take of marine mammals in such form as prescribed by the Assistant Administrator. All deaths or injuries to marine mammals occurring in the course of commercial fishing operations under the conditions of a general permit shall be immediately recorded in the log and reported in writing to the Regional Director, National Marine Fisheries Service, where a certificate application was made, or to an enforcement agent, or other designated agent of the National Marine Fisheries Service, at the earliest opportunity but no later than five days after such occurrence, except that if a vessel at sea returns to port later than five days after such occurrence, then it shall be reported within forty-eight hours after arrival in port. Reports must include:

(A) The location, time, and date of the death or injury;

(B) The identity and number of marine mammals killed or injured; and

(C) A description of the circumstances which led up to and caused the death or injury.

(vi) [Reserved]

(vii) The number of Dall's porpoise (*Phocoenoides dalli*) killed or injured by Japanese vessels operating in the U.S. EEZ is limited to an aggregate of 789 in the Bering Sea and 5250 in the North Pacific Ocean over the period 1987 to 1989, of which no more than 448 may be taken from the Bering Sea and no more than 2494 may be taken from the North Pacific Ocean in any single calendar year. The incidental take levels authorized by this subpart are reduced proportionately in the event that the Soviet Union reduces salmon quotas for 1988 or 1989 by more than 10 percent from the 1987 quota. Any permit issued under this part must indicate the measures by which the permit holder must comply with the conditions attached to the permit, and the reporting requirements of paragraph (d)(5)(v) of this section. Any permit issued under this part may allow retention of marine mammals for scientific purposes and will not require a separate permit under paragraph (d)(5)(iv) of this section.

(6) *Commercial passenger fishing vessels (CPFV).* (i) A certificate holder aboard the vessel may take marine mammals so long as the taking is limited to harassment and is an incidental occurrence in the course of the active sportfishing subject to the following restrictions (paragraphs (d)(6) (ii) through (vi)).

(ii) Takings are prohibited within 500 yards of a pinniped rookery or haul-out site.

(iii) A certificate holder aboard the CPFV must use only those non-lethal, non-injurious methods not including capture as approved in advance by the Assistant Administrator for Fisheries through publication in the FEDERAL REGISTER and stipulated in the General Permit for taking marine mammals.

(iv) Takings are allowed only while engaged in active sportfishing to prevent imminent marine mammal approaches to the vessel or to protect a passenger's catch or gear from depredation or damage, without inflicting death or injury to any marine mammal.

(v) All operators must ensure the safe use of the approved methods for preventing marine mammal sportfish-

National Marine Fisheries Service/NOAA, Commerce § 216.24

ing interaction and must satisfactorily complete such training as may be required by the Assistant Administrator for Fisheries.

(vi) All certificate holders must maintain records of incidental take of marine mammals in such form as prescribed by the Assistant Administrator for Fisheries. All incidents involving harassment of marine mammals must be immediately recorded and reported in writing to the Regional Director to whom the certificate application was made, or to an enforcement agent or other designated agent of the National Marine Fisheries Service, at the earliest opportunity, but no later than five days after such occurrence. At a minimum, reports must include:

(A) The time, date, and location of the taking;

(B) The type of harassment device used, and the number used at each occurrence;

(C) The number and species of affected marine mammals; and

(D) A description of any behavioral changes noted that may be due to using the harassment device.

(e) *Importation.* (1) It is illegal to import into the United States any fish, whether fresh, frozen, or otherwise prepared, if the fish have been caught with commercial fishing technology that results in the incidental kill or incidental serious injury of marine mammals in excess of that allowed under this part for U.S. fishermen or in excess of what is specified in subsection (e)(5) in the case of fishing for yellowfin tuna.

(2) The following fish and categories of fish, which the Assistant Administrator has determined may be involved with commercial fishing operations which cause the death or injury of marine mammals, are subject to the requirements of this section:

(i) *Yellowfin tuna.* The following U.S. Harmonized Tariff Schedule Item Numbers identify the categories of tuna and tuna products under which yellowfin tuna is imported into the United States and which are subject to the restrictions of paragraphs (e)(3) and (e)(5) of this section:

(A) Fish, fresh or chilled, excluding fish fillets and other fish meat of heading 0304:

0302.32.00.00.7 Yellowfin tunas.

(B) Fish, frozen, excluding fish fillets and other fish meat of heading 0304:

0303.42.00 Yellowfin tunas.
0303.42.00.20 Whole fish.
0303.42.00.40.6 Heads-on.
0303.42.00.40.60.1 Other.

(C) *Prepared fish:* Fish whole or in pieces, but not minced: Tuna, skipjack and Atlantic bonito:

1604.14.10.00.0 Tunas and skipjack; in airtight containers; in oil, (except cans marked as other than yellowfin tuna in a manner approved in advance by the National Marine Fisheries Service Southwest Regional Director).

1604.14.20.40.0 Tunas and skipjack in airtight containers; not in oil: * * * (except cans marked as other than yellowfin tuna in a manner approved in advance by the National Marine Fisheries Service Southwest Regional Director).

1604.14.30.40.8 Tunas and skipjack in airtight containers; other; (except cans marked as other than yellowfin tuna in a manner approved in advance by the National Marine Fisheries Service Southwest Regional Director).

1604.14.40.00.4 Tunas and skipjack; not in airtight containers; in bulk or in immediate containers weighing with their contents over 6.8 kg each, not in oil.

1604.14.50.00.1 Tunas and skipjack; not in airtight containers; other.

(ii) *Salmon and Halibut.* The following U.S. Harmonized Tariff Schedule Item Numbers identify the categories of salmon and halibut products which are imported into the United States and are subject to the restrictions of paragraph (e)(4) of this section:

(A) Fish, fresh or chilled, excluding fish fillets and other fish meat of heading 0302:

0302.12.00 Pacific salmon, Atlantic salmon, and Danube salmon.
0302.21.00.00.0 Halibut and Greenland turbot.

(B) Fish, frozen, excluding fish fillets and other fish meat of heading 0303:

0303.10.00 Pacific salmon.
0303.31.00.00.00.7 Halibut and Greenland turbot.

(C) Fish fillets and other fish meat, fresh, chilled or frozen:

0304.20.20.22.9 Frozen fillets; Halibut.
0304.20.60.50.5 Other; Halibut.

(D) Fish, dried, salted or in brine; smoked fish, whether or not cooked before or during the smoking process:

0305.41.00.00.3 Smoked fish, including fillets: Salmon.
0305.69.40.00.2 Fish salted but not dried or smoked and fish in brine: Salmon.

(E) Prepared or preserved fish; fish, whole or in pieces, but not minced:

1604.11 Salmon.

(3) *Yellowfin tuna.* (i) All shipments of tuna and tuna products listed in paragraph (e)(2)(i) of this section, from any nation, may be imported into the United States only if:

(A) accompanied by a separate Yellowfin Tuna Certificate of Origin (Standard Form 370-1); and

(B) in the case of a harvesting nation, the finding allowing importation is made as required by paragraph (e)(5)(i) of this section or, in the case of an intermediary nation, an embargo has not been imposed under paragraph (e)(5)(ix) of this section. Shipments of tuna and tuna products from nations which are both harvesting and intermediary nations must satisfy the provisions which apply to both categories of nations in order to be imported into the United States.

(ii) The Yellowfin Tuna Certificate of Origin must include the following information:

(A) Exporter (Name and Address);

(B) Consignee (Name and Address);

(C) Identity and quantity of the tuna to be imported, listed by U.S. tariff schedule number;

(D) Areas of harvest (ETP, western and central Pacific Ocean, Atlantic Ocean, Indian Ocean);

(E) Name of vessel(s) that caught the tuna, country under whose laws the vessel operated, and the date(s) of fishing trip of which the tuna was caught;

(F) Other documentation as may be required by the Assistant Administrator, subsequent to granting a finding in paragraph (e)(5) of this section;

(G) A declaration to be signed by either a responsible government official from the harvesting nation, the vessel master, the vessel owner's representative, or a representative of the cannery which processed the fish which states: "I certify that the above information is complete, true and cor-rect to the best of my knowledge and belief."

(4) *Salmon and Halibut.* All shipments of fish and fish products listed in paragraph (e)(2)(ii) of this section, from any nation, may not be imported into the United States unless the following conditions are met:

(i) The shipment is accompanied by a commercial invoice and/or a bill of lading indicating the following:

(A) Nation of registry of the fishing vessel(s) involved;

(B) Exporter (name and address);

(C) Consignee (name and address);

(D) Identity and quantity of the fish or fish products to be imported; and

(ii) The shipment is accompanied by a statement by a responsible official of the harvesting nation or the master of the vessel which caught the fish that such fish were not caught in a manner prohibited for U.S. fishermen by these regulations. The statement will identify the species, quantity, and exporter of the fish to which the statement refers, and be submitted at the time of importation; or

(iii) A responsible official of the harvesting nation may certify to the Assistant Administrator that all of its flag vessels are fishing in conformance with these regulations or that the fishing technology practiced by the harvesting nation with respect to the species of fish presented for importation into the United States does not result in deaths to marine mammals in excess of that which results from U.S. commercial fishing operations as prescribed by these regulations. Upon receipt of a statement of conformance, the Assistant administrator may then making a finding, and publish such finding in the FEDERAL REGISTER, that fish imports listed in paragraph (e)(2)(ii) of this section from the nation were not caught with commercial fishing technology which results in the incidental kill of marine mammals in excess of U.S. standards.

(5) *Yellowfin tuna.* (i) Any tuna or tuna products in the classifications listed in paragraph (e)(2)(i) of this section, from harvesting nations whose vessels of greater than 400 short tons carrying capacity operate in the ETP tuna purse seine fishery as determined by the Assistant Administrator, may

not be imported into the United States unless the Assistant Administrator makes an affirmative finding and publishes the finding in the Federal Register that:

(A) The government of the harvesting nation has adopted a regulatory program governing the incidental taking of marine mammals in the course of such harvesting that is comparable to the regulatory program of the United States; and

(B)' The average rate of incidental mortality by the vessels of the harvesting nation is comparable to the average rate of incidental mortality of marine mammals by U.S. vessels in the course of such harvesting as specified in paragraph (e)(5)(v)(E) and (e)(5)(v)(F) of this section.

(ii) A harvesting nation which desires an initial finding under these regulations that will allow it to import into the United States those products listed in paragraph (e)(2)(i) of this section must provide the Assistant Administrator with the following information:

(A) A detailed description of the nation's regulatory and enforcement program governing incidental taking of marine mammals in the purse seine fishery for yellowfin tuna, including:

(1) A description, with copies of relevant laws, implementing regulations and guidelines, of the gear and procedures required in the fishery to protect marine mammals, including but not limited to the following:

(i) A description of the methods used to identify problems and to take corrective actions to improve the performance of individual fishermen in reducing incidental mortality and serious injury. By 1990 the methods must identify individual operators with marine mammal mortality rates which are consistently and substantially higher than the majority of the nation's fleet, and provide for corrective training and, ultimately, suspension and removal from the fishery if the operator's performance does not improve to at least the performance of the majority of the fleet in a reasonable time period;

(ii) By 1990, a description of a regulatory system in operation which ensures that all marine mammal sets are completed through backdown to rolling the net to sack-up no later than one-half hour after sundown, except that individual operators may be exempted, if they have maintained consistently a rate of kill during their observed sundown sets which is not higher than that of the nation's fleet average during daylight sets made during the time period used for their comparability finding; and

(iii) By 1990, a description of its restrictions on the use of explosive devices in the purse seine fishery which are comparable to those of the United States.

(2) A detailed description of the method (e.g., Inter-American Tropical Tuna Commission (IATTC) or other international program observer records) and level of observer coverage by which the incidental mortality and serious injury of marine mammals will be monitored.

(B) A list of its vessels and any certified charter vessels of greater than 400 short tons carrying capacity which purse seined for yellowfin tuna at any time during the preceding year in the ETP, indicating the status of each such vessel during that period (i.e., actively fishing in ETP, fishing in other waters; in port for repairs; inactive) and the status of each vessel expected to operate in the ETP in the year in which the submission is made.

(C) A compilation of the best available data for each calendar year on the performance of any of its purse seine vessels (including certified charter vessels) fishing at any time for tuna associated with marine mammals within the ETP including the following:

(1) Total number of tons of yellowfin tuna observed caught in each fishing area by purse seine sets on:

(i) Common dolphin and

(ii) All other marine mammal species;

(2) Total number of marine mammals observed killed and the total number of marine mammals observed seriously injured in each fishing area by species/stock by purse seine sets on:

(i) Common dolphin and

(ii) All other marine mammal species;

(*3*) Total number of observed trips and total number of observed purse seine sets on marine mammals in each fishing area by the nation's purse seine fleet during the year;

(*4*) Total number of vessel trips and total number of purse seine sets on marine mammals in each fishing area by the nation's purse seine fleet during the year; and

(*5*) The total number of observed purse seine sets in each fishing area in which more than 15 marine mammals were killed.

(D) Data required by paragraph (e)(5)(ii)(C)(*2*) presented individually for the following marine mammal species/stocks: offshore spotted dolphin, coastal spotted dolphin, eastern spinner dolphin, whitebelly spinner dolphin, common dolphin, striped dolphin, and "other marine mammals".

(E) A description of the source of the data provided in accordance with paragraph (e)(5)(ii)(C) of this section. The observer program from which these data are provided must be operated by the IATTC or another international program in which the United States participates and must sample at least the same percentage of the fishing trips as the United States achieves over the same time period, unless the Assistant Administrator determines that an alternative observer program, including a lesser level of observer coverage, will provide a sufficiently reliable average rate of incidental taking of marine mammals for the nation.

(iii) A nation applying for its initial finding of comparability should apply at least 120 days before the desired effective date. The Assistant Administrator's determination on a nation's application for its initial finding will be announced and published in the FEDERAL REGISTER within 120 days of receipt of the information required in paragraph (e)(5)(ii) of this section.

(iv) A harvesting nation, which has in effect a positive finding under this section, may request renewal of its finding for the subsequent calendar year by providing the Assistant Administrator by July 31, or for nations participating in the IATTC observer program, within 14 days after the date the IATTC releases data on dolphin mortality, whichever is later, an update of the information listed in § 216.24(e)(5)(ii) which is current through the previous full calendar year.

(v) The Assistant Administrator will make an affirmative finding or renew an affirmative finding for a harvesting nation if:

(A) The harvesting nation has provided all information required by paragraphs (e)(5)(ii) and (e)(5)(iv) of this section;

(B) The nation's regulatory program is comparable to the regulatory program of the United States as described in paragraphs (a), (b), (c), (d)(2), and (f) of this section and the nation has incorporated into its regulatory program such additional prohibitions as the United States may apply to its own vessels within 180 days after the prohibition applies to U.S. vessels, except that provisions for identifying and removing problem operators, prohibiting sundown sets, and restricting the use of explosives described in (e)(5)(ii)(A)(*1*) of this section must be applied by a nation to its vessels by January 1, 1990;

(C) The data on marine mammal mortality and serious injury submitted by the harvesting nation are determined to be accurate;

(D) The observer coverage of fishing trips was equal to that achieved by the United States during the same time period or, if less, was determined by the Assistant Administrator to provide a sufficiently accurate sample of the nation's fleet mortality rate;

(E) For findings using data collected after 1988, the average kill-per-set rate for the longest period of time for which data are available, up to 5 consecutive years, or for the most recent year, whichever is lower, is no more than 25 percent greater than the U.S. average for the same time period, after the U.S. mortality rate is weighted to account for dissimilar amounts of fishing effort between the two nations in the three ETP fishing areas and for common dolphin and other marine mammal species, except as provided in paragraph (e)(5)(v)(F) of this section for findings made in 1990;

(F) Based on data collected through the 1989 fishing year, a nation's average mortality rate is no more than twice that of the U.S.-flag fleet for the same period, after the U.S. mortality rate is weighted to account for dissimilar amounts of fishing effort between the two nations in the three ETP fishing areas and for common dolphin and other marine mammal species;

(G) For the 1989 fishing year and subsequent years, the nation's observed kill of eastern spinner dolphin (*Stenella longirostris*) and coastal spotted dolphin (*Stenella attenuata*) is no greater than 15 percent and 2 percent, respectively, of the nation's total annual observed dolphin mortality; and

(H) The nation has complied with all reasonable requests by the Assistant Administrator for cooperation in carrying out dolphin population assessments in the ETP.

(vi) The Assistant Administrator may require verification of statements made in connection with requests to allow importations.

(vii) A finding is valid only for the period for which it was issued and may be terminated before the end of the year if the Assistant Administrator finds that the nation no longer has a comparable regulatory program or kill rate.

(viii) The Assistant Administrator may reconsider a finding upon a request from and the submission of additional information by the harvesting nation, if the information indicates that the nation has met the requirements under paragraph (e)(5)(v) of this section. For a harvesting nation whose marine mammal mortality rate was found to exceed the acceptable levels prescribed in paragraphs (e)(5)(v)(E), (e)(5)(v)(F), or (e)(5)(v)(G) of this section, the additional information must include data collected by an acceptable observer program which demonstrate that the nation's fleet marine mammal mortality rate improved to the acceptable level during the period including at a minimum January 1 through June 30 of the year following the year in which the nation's mortality rate was found to exceed acceptable levels.

(ix) Any tuna or tuna products in the classifications listed in paragraph (e)(2)(i) of this section, from any intermediary nation, may not be imported into the United States if the Assistant Administrator determines and publishes the determination in the FEDERAL REGISTER that the intermediary nation has not provided reasonable proof and has not certified to the United States that it has acted to ban the importation into its nation of all yellowfin tuna and tuna products from those nations from which the United States has banned the importation of yellowfin tuna and tuna products. The intermediary nation's ban must be effective within sixty days of the effective date of the U.S. ban, and the intermediary nation must supply the Assistant Administrator with the reasonable proof and certification within 90 days of the effective date of the U.S. ban in order to avoid imposition of the embargo on the 91st day. The Secretary will not require an intermediary nation to certify that it has implemented a ban on yellowfin tuna and tuna products from a nation embargoed by the United States, if the Secretary is satisfied that the intermediary nation continues to derive its tuna products, directly or indirectly, from sources other than that embargoed nation. An embargo under this paragraph may be lifted by the Assistant Administrator upon a determination announced in the FEDERAL REGISTER, based upon new information supplied by the embargoed nation, that the nation has acted to ban yellowfin tuna and tuna products from those nations from which the United States has banned the importation of yellowfin tuna and tuna products.

(x) After six months of an embargo being in place against a nation under this section, that fact shall be certified to the President for purposes of certification under section 8(a) of the Fishermen's Protective Act of 1967 (22 U.S.C. 1978(a)) for as long as the embargo is in effect.

(xi) The Assistant Administrator will promptly advise the Department of State of embargo decisions and actions and of finding determinations.

(6) *Fish refused entry*. If fish is denied entry under the provisions of § 216.24(e)(3) or (e)(4), the District Director of Customs shall refuse to release the fish for entry into the United States and shall issue a notice

of such refusal to the importer or consignee.

(7) *Release under bond. Provided however,* That fish not accompanied or covered by the required documentation or certification when offered for entry may be entered into the United States if the importer or consignee gives a bond on Customs Form 7551, 7553, or 7595 for the production of the required documentation or certification. The bond shall be in the amount required under 19 CFR 25.4(a). Within 90 days after such Customs entry, or such additional period as the District Director of Customs may allow for good cause shown, the importer or consignee shall deliver a copy of the required documentation or certification to the District Director of Customs, and an original of the required documentation or a copy of the certification to the Regional Director of the National Marine Fisheries Service, unless the District Director of Customs has received notification from the National Marine Fisheries Service that the fish is covered by a certification. If such documentation, certification, or notification is not delivered to the District Director of Customs for the port of entry of such fish within 90 days of the date of Customs entry or such additional period as may have been allowed by the District Director of Customs for good cause shown, the importer or consignee shall redeliver or cause to be redelivered to the District Director of Customs those fish which were released in accordance with this paragraph. In the event that any such fish is not redelivered within 30 days following the date specified in the preceding sentence, liquidated damages shall be assessed in the full amount of bond given on Form 7551. When the transaction has been charged against a bond given on Form 7553 or 7595, liquidated damages shall be assessed in the amount that would have been demanded under the preceding sentence under a bond given on Form 7551. Fish released for entry into the United States through use of the bonding procedure provided in this paragraph shall be subject to the civil and criminal penalties and the forfeiture provisions provided for under the Act if:

(i) The required documentation or certification is not delivered to the Regional Director of the National Marine Fisheries Service within 90 days of the date of Customs entry, or such additional period as may have been allowed by the District Director of Customs for good cause shown, or

(ii) The required certification is not on file in the office of the Assistant Administrator, National Marine Fisheries Service, National Oceanic and Atomspheric Administration, Department of Commerce, Washington, DC 20235, within this 90 day period or such additional period as may have been allowed by the District Director of Customs for good cause shown. Fish refused entry into the United States shall also be subject to the civil and criminal penalties and the forfeiture provisions provided for under the Act.

(8) *Disposition of fish refused entry into the United States; redelivered fish.* Fish which is denied entery under § 216.24(e)(3) or (e)(4) or which is redelivered in accordance with § 216.24(e)(7) and which is not exported under Customs supervision within 90 days from the date of notice of refusal of admission or date of redelivery shall be disposed of under Customs laws and regulations. *Provided however,* That any disposition shall not result in an introduction into the United States of fish caught in violation of the Marine Mammal Protection Act of 1972.

(9)(i) *Restrictions on yellowfin and bigeye tuna imports.* No yellowfin or bigeye tuna harvested by a purse seine vessel fishing on porpoise in the eastern tropical Pacific Ocean (the "ETP") during the period that such fishing is prohibited for vessels of the United States under § 216.24(a)(4)(i) in 1986, may be imported into the United States. Any tuna harvested in the ETP before the closure date must be placed in a bonded warehouse, aboard a common carrier, or aboard a vessel in port awaiting to unload before the closure date in order to be imported into the United States after the closure date.

(ii) *Definitions for purposes of § 216.24(e)(9).* (A) "Closure date" means the date on which the allowable quota on incidental mortality permitted under the general permit will be reached as announced under the provisions of § 216.24(d)(2)(i)(B).

(B) "Closure period" means a period beginning on the closure date and ending on December 31, 1986, during which United States purse seine vessels are prohibited from catching tuna by setting on porpoise.

(C) "ETP" means the eastern tropical Pacific Ocean which includes the Pacific Ocean area bounded by 40° N. latitude, 40° S. latitude, 160° W. longitude, and the coastline of North, Central, and South America.

(D) "NMFS" means the National Marine Fisheries Services, National Oceanic and Atmospheric Administration, Department of Commerce.

(iii) *Authorization required for yellowfin and bigeye tuna imports.* Any yellowfin or bigeye tuna harvested by a vessel of a country for which the Assistant Administrator has made a finding under paragraph (e)(5)(i) of this section, that is offered for importation between the closure date and June 30, 1987, must be accompanied by a letter of authorization from NMFS.

(iv) *Procedure to obtain letter of authorization.* Any person desiring to obtain a letter of authorization from NMFS to allow the importation of yellowfin or bigeye tuna must submit to the Regional Director, NMFS, 300 S. Ferry Street, Terminal Island, California 90731, a declaration signed by a responsible government official of the country whose flag vessel caught the tuna, that the tuna being offered for importation was not taken by fishing on porpoise during the closure period. The declaration of the responsible government official must include the information listed in either paragraph (e)(9)(iv) (A), (B), (C), (D), or (E) of this section, as appropriate.

(A) If the tuna to be imported was taken on a fishing trip any part of which was in the ETP during the closure period, the declaration must include the following information:

(1) The name of the vessel(s) which harvested the tuna,

(2) The date of the trip(s) on which the tuna was harvested,

(3) A statement that the vessel carried an observer approved by the government on every portion of the trip that occurred after the closure date, and

(4) A statement that the observer certified that no tuna was harvested by fishing on porpoise after the closure date.

(B) If the tuna to be imported was taken on a fishing trip in the ETP that ended before the closure date, the declaration must include the following information:

(1) The names of the vessel(s) which harvested the tuna,

(2) The dates of the trip(s) on which the tuna was harvested, and one of the following sets of information:

(i) The date the tuna arrived in port, the date when the tuna was placed in a bonded warehouse, the date the tuna was removed from the bonded warehouse, and the name and address of the bonded warehouse. Documentation establishing that the tuna was in port on the closure date and documentation from the bonded warehouse that shows the dates that the tuna was placed into the warehouse and taken out of the warehouse must accompany the declaration.

(ii) The date the tuna arrived in port, the date when the tuna was placed on a common carrier for shipping, and the name and address of the common carrier for shipping, and the name and address of the common carrier. Documentation establishing that the tuna was in port on the closure date and documentation from the common carrier that shows the date(s) that the tuna was placed aboard the carrier must accompany the declaration.

(iii) The date the tuna arrived in port, awaiting unloading, the date the vessel is schedule for unloading, the intended disposition of the tuna (i.e., common carrier, bonded warehouse, offloading in Puerto Rico, Venezuela, etc.) and documentation from a port official verifying the arrival date of the vessel.

(C) If the tuna to be imported was taken by fishing on porpoise on a fishing trip that began after the closure period ends, the declaration must include the following information:

(1) The names of the vessel(s) which harvested the tuna,

(2) The dates of the trip(s) on which the tuna was harvested, and

(3) The location of the harvest.

(D) If the tuna to be imported was taken on a fishing trip on which the vessel was not in the ETP for any portion of the trip, the declaration must include the following information:

(1) The names of the vessel(s) which harvested the tuna,

(2) The dates of the trip(s) on which the tuna was harvested, and

(3) The areas in which the vessel(s) was during the trip(s) (e.g., the western Pacific Ocean).

(E) If the tuna to be imported was taken in part or entirely by tuna purse seine vessels of 400 short tons carrying capacity or less, (and any remaining tuna was not caught by tuna purse seine vessels of larger than 400 short tons) the declaration must include the following information:

(1) The name(s) of the vessel(s) which caught the tuna;

(2) The carrying capacity in short tons of the(se) vessel(s); and

(3) the quantity of tuna being imported that was caught by the(se) vessel(s).

(v) *Disposition of tuna not accompanied by required documentation.* (A) Tuna that requires a letter of authorization under paragraph (e)(9)(iii) of this section that is offered for importation without the required letter of authorization must be either—

(1) Exported under Customs supervision within 60 days,

(2) Placed into a bonded warehouse, or

(3) Disposed of under Customs laws and regulations, as long as that disposition does not result in its introduction into the United States.

(B) The importer will remain liable for any expenses incurred in the storage and/or disposal of tuna refused admission under these regulations. If, within 60 days of fish being placed into a bonded warehouse, the District Director of Customs receives appropriate documentation for that fish, the fish will be allowed to be entered into the United States, otherwise it will be disposed of as set forth in paragraph (e)(9)(v) (A) of (C) of this section.

(f) *Observers.* (1) The vessel certificate holder of any certificated vessel shall, upon the proper notification by the National Marine Fisheries Service, allow an observer duly authorized by the Secretary to accompany the vessel on any or all regular fishing trips for the purpose of conducting research and observing operations, including collecting information which may be used in civil or criminal penalty proceedings, forfeiture actions, or permit or certificate sanctions.

(2) Research and observation duties shall be carried out in such a manner as to minimize interference with commercial fishing operations. The navigator shall provide true vessel locations by latitude and longitude, accurate to the nearest minute, upon request by the observer. No owner, master, operator, or crew member of a certificated vessel shall impair or in any way interfere with the research or observations being carried out.

(3) Marine mammals killed during fishing operations which are accessible to crewmen and requested from the certificate holder or master by the observer shall be brougnt aboard the vessel and retained for biological processing, until released by the observer for return to the ocean. Whole marine mammals designated as biological specimens by the observer shall be retained in cold storage aboard the vessel until retrieved by authorized personnel of the National Marine Fisheries Service when the vessel returns to port for unloading.

(4) The Secretary shall provide for the payment of all reasonable costs directly related to the quartering and maintaining of such observers on board such vessels. A vessel certificate holder who has been notified that the vessel is required to carry an observer, via certified letter from the National Marine Fisheries Service, shall notify the office from which the letter was received at least five days in advance of the fishing voyage to facilitate observer placement. A vessel certificate holder who has failed to comply with the provisions of this section may not engage in fishing operations for which a general permit is required.

(5) It is unlawful for any person to forcibly assault, impede, intimidate, interfere with, or to influence or attempt to influence an observer, or to

harass (including sexual harassment) an observer by conduct which has the purpose or effect of unreasonably interfering with the observer's work performance, or which creates an intimidating, hostile, or offensive environment. In determining whether conduct constitutes harassment, the totality of the circumstances, including the nature of the conduct and the context in which it occurred, will be considered. The determination of the legality of ·a particular action will be made from the facts on a case-by-case basis.

(6)(i) All observers must be provided sleeping, toilet and eating accommodations at least equal to that provided to a full crew member. A mattress or futon on the floor or a cot is not acceptable in place of a regular bunk. Meal and other galley privileges must be the same for the observer as for other crew members.

(ii) Female observers on a vessel with an all-male crew must be accommodated either in a single-person cabin or, if reasonable privacy can be ensured by installing a curtain or other temporary divider, in a two-person cabin shared with a licensed officer of the vessel. If the cabin assigned to a female observer does not have its own toilet and shower facilities that can be provided for the exclusive use of the observer, then a schedule for time-sharing common facilities must be established before the placement meeting and approved by NMFS and must be followed during the entire trip.

(iii) In the event there are one or more female crew members, the female observer may be provided a bunk in a cabin shared solely with female crew members, and provided toilet and shower facilities shared solely with these female crew members.

(7)(i) A vessel certificate of inclusion holder (or vessel owner in the case of a new application) may seek an exemption from carrying a female observer on a vessel by applying to the Southwest Regional Director when applying for the vessel certificate of inclusion until July 10, 1989 and establishing the following:

(A) The vessel will have an all-male crew;

(B) The vessel has fewer than two private (one-person) and semi-private (two-person) cabins in total (excluding the captain's cabin);

(C) A temporary divider like a curtain cannot be installed in the private or semi-private cabin (excluding the captain's cabin) to provide reasonable privacy; and

(D) There are no other areas (excluding the captain's cabin) that can be converted to a sleeping room without either significant expense or significant sacrifice to the crew's quarters.

(ii) The exclusion criteria in paragraph (f)(7)(i) of this section can be met without having to provide the captain's cabin for the observer. The application for an exemption must also include an accurate diagram of the vessel's living areas, and other areas possibly suitable for sleeping. Additional documentation to support the application may also be required, as may an inspection of the vessel. The exemption, once granted, is valid for the same calendar year as the vessel certificate of inclusion, and the exemption must be renewed annually to remain valid. The vessel certificate of inclusion holder is responsible for reporting to the Regional Director any changes aboard the vessel within 15 days of the change which might affect the continued eligibility for an exemption. The Southwest Regional Director will revoke an exemption if the criteria for an exemption are no longer met.

(g) *Penalties and rewards:* Any person or vessel subject to the jurisdiction of the United States shall be subject to the penalties provided for under the Act for the conduct of fishing operations in violation of these regulations. The Secretary shall recommend to the Secretary of the Treasury that an amount equal to one-half of the fine incurred but not to exceed $2,500 be paid to any person who furnishes information which leads to a conviction for a violation of these regulations. Any officer, employee, or designated agent of the United States or of any State or local government who furnishes information or renders service in the performance of his official duties shall not be eligible for payment under this section.

[45 FR 72187, Oct. 31, 1980]

EDITORIAL NOTE: For Federal Register citations affecting § 216.24, see the List of CFR Sections Affected in the Finding Aids section of this volume.

EFFECTIVE DATE NOTE: At 54 FR 415, January 6, 1989, § 216.24 was amended by revising paragraph (d)(2)(vii)(C) and adding a new paragraph (d)(2)(vii)(E), effective January 1, 1989, except the amendments to § 216.24(d)(2)(vii)(C)(1) and (5), and (viii), which contain information collection requirements and will not be effective until approved by the Office of Management and Budget. When approval is obtained, NOAA will publish a notice of effective date for these paragraphs in the Federal Register.

§ 216.25 Exempted marine mammals and marine mammal products.

(a) The provisions of the Act and these regulations shall not apply:

(1) To any marine mammal taken before December 21, 1972, or

(2) To any marine mammal product if the marine mammal portion of such product consists solely of a marine mammal taken before such date.

(b) The prohibitions contained in § 216.12(c) (3) and (4) shall not apply to marine mammals or marine mammal products imported into the United States before the date on which a notice is published in the FEDERAL REGISTER with respect to the designation of the species or stock concerned as depleted or endangered.

(c) Section 216.12(b) shall not apply to articles imported into the United States before the effective date of the foreign law making the taking or sale, as the case may be, of such marine mammals or marine mammal products unlawful.

§ 216.26 Collection of certain marine mammal parts.

(a) Any bones, teeth or ivory of any dead marine mammal may be collected from a beach or from land within ¼ of a mile of the ocean. The term "ocean" includes bays and estuaries.

(b) Marine mammal parts so collected may be retained if registered within 30 days with an agent of the National Marine Fisheries Service, or an agent of the Bureau of Sport Fisheries and Wildlife.

(c) Registration shall include (1) the name of the owner, (2) a description of the article to be registered and (3) the date and location of collection.

(d) Title to any marine mammal parts collected under this section is not transferable unless consented to, in writing, by the Secretary.

APPENDIX

2

Forms and data format for documenting flotsam locations and characteristics. The Inter-American Tropical Tuna Commission requires that all observers in its international program keep these records on every cruise.

Comisión Interamericana del Atún Tropical
REGISTRO DE OBSERVACIONES DE OBJETOS FLOTANTES

Crucero No: _____ Registro No: _____ Objeto flotante marcado por la CIAT? (S/N) _____ No. _____

Técnico científico: _____ Nombre del Barco: _____

Lance No. _____ Fecha: ___ ___ ___ ___ ___ ___ Hora del lance/avistamiento _____
(AA/MM/DD)

Latitud: ___ ___ ___ ___ N/S Longitud: ___ ___ ___ ___ ___ W Temperatura del agua ___ ___ . ___ ° C / °F

Nubosidad en lance/avistamiento: _____ ¿Palo fuera de la red ? (S/N) _____ Distancia _____
(use Tabla de Código 3 del ID)

Claridad del agua (señale una)   Clara   Turbia   Muy turbia   ¿Corriente fuerte ? (S/N) _____

INFORMACION SOBRE EL OBJETO FLOTANTE

¿Lances o avistamientos previos sobre este palo ? (S/N) _____ Nos. de registro previos : _____ - _____

_____ - _____ - _____ - _____ - _____ - _____ - _____ - _____ - _____ - _____ - _____ - _____

OBJETO FLOTANTE   Tipo : _____   Forma : _____   Material: _____   Color: _____
(use las Tablas de Códigos 10 y 11)

Describa : (incluya dimensiones y características importantes) _____

_____

_____

_____

% debajo del agua _____ [   ]   Tiempo estimado en el agua :   Poco   Bastante   Mucho
                             prf   (señale uno)

Si es un árbol o planta acuática: ¿Puede identificarlo ? _____   Cortado (S/N) _____
                                                                    (marcas de machete, sierra o hacha)

Raíces (S/N) _____   Ramas (S/N) _____   Corteza (S/N) _____   Hojas (S/N) _____

Posible origen del objeto flotante _____

_____

¿Otros objetos flotantes en el área? (S/N) _____ Tipo _____ ¿Señales de pesca previa sobre este

objeto ? (Cabos, redes, radiotransmisores, lances de otros barcos) _____

_____

¿Muestra tomada ? (S/N) _____ Muestra No. _____

CIAT forma ROF v. 1.0   6/89

INFORMACION SOBRE FAUNA Y FLORA

CAPTURA DE ATUN (use la Tabla de Códigos 2)

	Especies capturadas	Toneladas	Rango de peso *
1.			
2.			
3.			
4.			

* Indique el rango aproximado de peso para cada especie de atún capturada. Especifique si usa lbs. ó kgs.

OTRA FAUNA ASOCIADA CON EL OBJETO FLOTANTE (use las Tablas de Códigos 2 y 12)

	Especies	Estimación * de números **o** toneladas		Estimación* del rango de peso (lb/kg) **o** rango de longitud (pies/m)	
1.					
2.					
3.					
4.					
5.					
6.					
7.					

* Si no puede estimar, indique mucho/grande, medio, o poco/chico

PAJAROS (use la Tabla de Códigos 13)

	Especies avistadas	No. estimado*
1.		
2.		
3.		
4.		

FAUNA/FLORA EN EL OBJETO FLOTANTE
(use la Tabla de Códigos 13)

1.		
2.		
3.		% cubierto con
4.		epibiota ____

DIBUJO DEL OBJETO FLOTANTE (incluya dimensiones, áreas cubiertas con epibiota, y ángulo de flotación; vea los ejemplos)

vista desde ARRIBA	vista de LADO

Comentarios adicionales: ☐ 1   ☐ 2/d   ☐ 3   ☐ ang

Inter-American Tropical Tuna Commission
FLOTSAM INFORMATION RECORD

Cruise No. : _____ Record No.: _____ IATTC tagged flotsam? (Y/N) _____ No. _____

Technician name: _____ Vessel name: _____

Set No. _____ Date: ___ ___ ___ ___ ___ ___ Let go / sighting time _____
(YY/MM/DD)

Latitude ___ ___ ___ ___ N/S Longitude ___ ___ ___ ___ ___ W Water temperature ___ ___ . ___ ° C / °F

Cloud cover at let go / sighting: _____ Log outside the net ? (Y/N) _____ How far ? _____
(use Code Table 3)

Water clarity Clear Turbid Very turbid Beaufort No. at
(circle one) let go / sighting _____ Strong current ? (Y/N) _____
(use Code Table 7)

FLOTSAM INFORMATION

Previous sets or sightings of this log ? (Y/N) _____ Previous record Nos. : _____ - _____ - _____

_____ - _____ - _____ - _____ - _____ - _____ - _____ - _____ - _____ - _____ - _____ - _____

FLOTSAM Type : _____ Shape : _____ Material: _____ Color: _____
(use Code Tables 10 and 11)

Describe : (include <u>dimensions</u> and distinctive characteristics) _____

_____

_____

_____

% under water _____ [____] dpt Estimated time in water : (circle one) Short Medium Long

If tree or aquatic plant: Can you identify ? _____ Cut (machete, axe, or saw marks) (Y/N) _____

Roots (Y/N) _____ Branches (Y/N) _____ Bark (Y/N) _____ Leaves (Y/N) _____

Possible source of flotsam _____

_____

Other flotsam in area? (Y/N) _____ Type _____ Any sign of previous fishing activity on flotsam ? (Net

twine, webbing, beepers, known sets by other boats) _____

_____

Sample taken (Y/N) _____ Sample No. _____

IATTC FIR form v. 1.0 6/89

FAUNA AND FLORA INFORMATION

TUNA CATCH   (use Code Table 2)

	Species caught	Tonnage	Weight range *
1.	_____	_____	_____
2.	_____	_____	_____
3.	_____	_____	_____
4.	_____	_____	_____

       * Give approximate weight range for each tuna species caught. Specify if lbs. or kgs.

OTHER FAUNA ASSOCIATED WITH FLOTSAM   (use Code Tables 2 and 12)

	Species	Estimated * numbers	**or**	tonnage	Estimated * weight range (lb/kg)	**or**	length range (ft/m)
1.	_____	_____		_____	_____		_____
2.	_____	_____		_____	_____		_____
3.	_____	_____		_____	_____		_____
4.	_____	_____		_____	_____		_____
5.	_____	_____		_____	_____		_____
6.	_____	_____		_____	_____		_____
7.	_____	_____		_____	_____		_____

       * If cannot estimate indicate high/large, medium, or low/small

BIRDS (use Code Table 13)

	Species sighted	* Estimated No.
1.	_____	_____
2.	_____	_____
3.	_____	_____
4.	_____	_____

FAUNA/FLORA ATTACHED TO FLOTSAM (use Code Table 13)

1.	_____	
2.	_____	
3.	_____	% covered with
4.	_____	epibiota _____

FLOTSAM DRAWING (include dimensions, areas covered with epibiota, and floating angle; see examples)

TOP view	SIDE view

   [ ]   [ ]   [ ]   [ ]

Additional notes :   1   2/d   3   ang

Code Table 2 - Tuna Species

Species	Letter Code		Number Code
	Span	Eng	
Yellowfin	AA	YF	110
Skipjack	BA	SJ	111
Bigeye	OG	BE	106
Bluefin	AZ	BF	101
Albacore	ALB	ALB	102
Black skipjack	BN	BSJ	103
Bullets	MEL	BUL	104
Bonito	BON	BON	105

Code Table 11 - Flotsam Shape

Shape	Letter Code		Number Code
	Span	Eng	
Cylindrical	CILN	CYLD	1
Polygonal/box-like	POLI	POLY	2
Rounded	REDN	ROUN	3
Irregular	IRRG	IRRG	4
Aggregated	AGRE	AGGR	5
Other shape	OTFO	OTSH	6

Code Table 14 - Colors

COLORS

Red	Rojo	1
Green	Verde	2
Orange	Naran	3
Blue	Azul	4
Yellow	Amar	5
Black	Negro	6
White	Blanc	7
Brown	Cafe	8
Silver	Plata	9

Code Table 10 - Flotsam Type

	Letter Code		Number Code
	Span	Eng	

Wooden objects

Natural

	Span	Eng	No.
Palm	PALM	PALM	11
Banana	PLAT	BANA	12
Bamboo	BMBU	BAMB	13
Mangrove	MNGL	MANG	14
Cane	CANA	CANE	15
Hay	PAJA	HAY	16
Fruits	FRUT	FRUT	17
Unidentified tree	PALO	UNTR	10

Man-made

	Span	Eng	No.
Boats or parts of boats	BARP	BOTP	21
Pallets/crates	CATA	PACR	22
Planks/boards	TAPO	PLBR	23
Plywood	TRIP	PLYW	24
Rafts	BALS	RAFT	25
Spools	CARR	SPOL	26
Wooden drums/buckets	TACU	WDRB	27

Other objects

Natural

	Span	Eng	No.
Dead whale	BAMU	DEWH	31
Other dead animal	OAMU	ODAN	32
Kelp patty	ALGA	KELP	33

Man-made

	Span	Eng	No.
Rope	SOGA	ROPE	41
Fishing gear	EPES	FSGR	42
Buoy	BOYA	BUOY	43
Life preservers	SALV	LIPR	44
Other discarded equipment	OEDS	ODIE	45
Tires	LLAN	TIRE	46
Foam	HUES	FOAM	47
Plastic drums	TAPL	PDRU	48
Other plastic objects	OPLA	OPOB	49
Trash	BASU	TRSH	50
Research buoy	BOYI	RBUY	51
Fish-Aggregating Device	FADS	FADS	52
Other objects	OTOB	OTOB	59
Unidentified objects	OBNI	UNOB	40

Code Table 12 - Fauna Under Flotsam

Species	Letter Code Span	Letter Code Eng	Number Code
Billfish			
Marlin	MARL	MARL	121
Sailfish	PVEL	SF	122
Swordfish	PESP	SWF	123
Other billfish	OTPI	OTBF	120
Other medium/large fish			
Dorado/Mahi mahi	DORA	DOR	131
Wahoo	PETO	WA	132
Rainbow runners	SALM	RRUN	133
Yellowtail	JUAA	YT	134
Other large fish	OTPG	OTLF	130
Other fish			
Triggerfish	CCHI	TRGF	141
Small baitfish	CAPE	SMBT	142
Other small fish	OTPP	OTSF	140
Sharks and rays			
Blacktip shark	TIPN	BTSH	151
Whitetip shark	TIPB	WTSH	152
Hammerhead	TIMA	HHS	153
Other shark	OTIB	OTSH	150
Unidentified shark	TINI	SHARK	154
Manta ray	MANT	MANT	155
Stingray	RAYA	STRY	156
Others			
Marine mammals see Code Table 9			
Sea turtles	TOMA	TURT	161
Invertebrates	INVE	INVB	162
Other fauna	OTFA	OTFA	172
Unidentified fish	PENI	UNFS	170
None (no fauna)	NFAU	NOFA	174

Code Table 13 - Birds and Epibiota

Name	Letter Code Span	Letter Code Eng	Number Code

Birds

Boobies
Red-footed ........................BOPR.......RFBO..................21
Masked ...........................BOMA......MABO..................22
Brown .............................BOCA......BRBO..................23
Unidentified booby .............BONI.......UNBO..................20

Shearwaters
Wedge-tailed ....................PRGA.....WTSW..................31
Small (Manx, Audubon) .......PRPE.....SMSW..................32
Pink-footed .......................PRRO......PFSW..................33
Unidentified shearwater .......PRNI.......UNSW..................30

Terns
Black and white .................GOBN.......BWTE..................41
White ..............................GOBL.......WHTE..................42

Frigate/Man of War ...............FREG........MOW..................51
Petrels ...............................PETR.......PETR..................60
Jaegers ..............................ESTR........JAEG..................70

Other bird ...........................OTPJ.......OTBD..................81
Unidentified bird .....................PJNI.......UNBD..................80

Epibiota

Acorn barnacles .....................BALA......ACBR..................1
Gooseneck barnacles .............PERB.......GNBR..................2
Crabs ..................................CANG......CRAB..................3
Green seaweed .......................ALVE......GRSW..................4
Other seaweed .......................OTAL......OTSW..................5
Limpets ...............................LAPA........LIMP..................6
Chitons ...............................QUIT......CHTN..................7
Mussels ...............................ALME......MUSS..................8
Guano .................................GUAN......GUAN..................9
Sea turtle ............................TOMA.......TURT..................11
Other epibiota .......................OTEB.......OTEB..................10
Unidentified epibiota .............EBNI.......UNEB..................99

APPENDIX

3

Notice sent by the Inter-American Tropical Tuna Commission to tuna-seiner owners and operators in 1987 asking for their cooperation in tracking tagged flotsam.

NOTICE TO FISHERMEN
Tagged Drifting Objects

The Inter-American Tropical Tuna Commission (IATTC) will be tagging drifting objects, such as tree trunks, branches, etc., off the Colombian coast during August and September of 1987. The IATTC will also attempt to tag tunas associated with these objects.

The purpose of this project is to monitor the movement of drifting objects in the eastern Pacific tuna fishing grounds and to study the association of tunas with these objects.

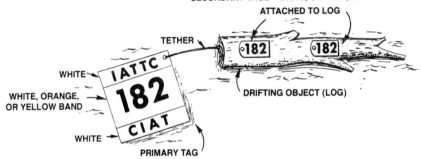

Primary tag: One plastic floating square, 2 ft × 2 ft × 3/16 in, tethered to the object.

Secondary tag(s): One or more colored plastic tags, 6 in × 4 in, attached directly to the drifting object.

If you see any of these tagged drifting objects, please:

1. Record in your logbook and notify one of the offices on the back of this page, the following information:
   a. Tag number (or color if number cannot be seen), and the condition of the tags
   b. When and where it was sighted
   c. Abundance of fish around it
   d. Condition of the drifting object
2. Do not remove the tags from the log, as there is no reward for them.

Inter-American Tropical Tuna Commission

Index